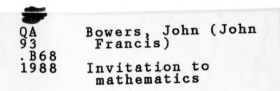

QA
93
.B68
1988

Bowers, John (John Francis)

Invitation to mathematics

$39.95

DATE		

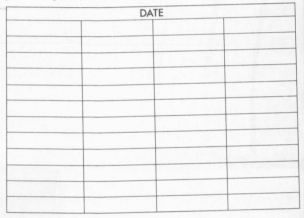

Invitation to
MATHEMATICS

INVITATION SERIES

Invitation to
MATHEMATICS

John Bowers

BASIL BLACKWELL

Copyright © John Bowers 1988

First published 1988

Basil Blackwell Ltd
108 Cowley Road, Oxford, OX4 1JF, UK

Basil Blackwell Inc.
432 Park Avenue South, Suite 1503
New York, NY 10016, USA

British Library Cataloguing in Publication Data

Bowers, John
Invitation to mathematics.
1. Mathematics—1961–
I. Title
510   QA37.2

ISBN 0-631-14641-5
ISBN 0-631-14642-3 Pbk

Library of Congress Cataloging in Publication Data

Bowers, John (John Francis)
Invitation to mathematics.

Includes index.
1. Mathematics—Popular works. 2. Mathematics—Philosophy. 3. Mathematics—History. I. Title.
QA93.B68 1988   510   87-29902
ISBN 0-631-14641-5
ISBN 0-631-14642-3 (pbk.)

Typset in Bembo 10 on 12pt
by Columns of Reading
Printed in Great Britain by Page Bros. Ltd, Norwich

Magistra philosophiae,
undecim solidis nouem denariisque solutis,
risisti pariter omnes uictorias cladesque,
Clothilda,
et super margaritis augustisque deligeris:
tibi hic liber dedicatus est.

Contents

Contents

Preface

Everybody knows some mathematics, therefore issuing an invitation to mathematics is correspondingly easier than issuing one to, say, Premonstratensian Ichthyography, though the mathematics known to a majority is so uninviting that it suggests that an invitation is more urgently needed here than for any other subject. Of course, it is true that some elementary mathematics is not unduly exciting, but some of the more advanced parts are of absorbing interest, and mathematics eventually reaches some astonishing conclusions, just as physics does. However, the really remarkable aspect of these results is that they are the inescapable, proven truth; but we shall not spoil them by considering them here, because this is a book *about* mathematics, not a mathematics book, such as a textbook. Naturally, some mathematics is included by way of illustration, particularly in discussions of mathematical methods.

This invitation is extended to everyone who wishes to know more about mathematics, and it is certainly not directed only to those 'brilliant mathematicians' whose speed of thought far outdistanced their classmates in the early arithmetic classes. Indeed, as I shall indicate in the last chapter, many of those 'brilliant mathematicians' were not mathematicians at all, while some of the comparatively despised majority of the class had genuine mathematical gifts. In order to address all who can profit by this book, the mathematical background required for reading it is suitably elementary. It is certainly enough to know how to solve quadratic and simultaneous linear equations and to know about Pythagoras' Theorem. Actually, it is not obvious that this amount of knowledge is necessary for reading this book, but the list serves to show what kind of mathematical knowledge is required. This is not to say that this book does not include some mathematically difficult parts, because otherwise we could not discuss mathematical analysis. However, the level of difficulty has been kept down by restricting the mathematical discussions to topics belonging to first-year university courses. On the other hand, the first and last chapters of the book contain hardly any mathematics, but are concerned, respectively, with the nature and extent of mathematics and the nature and intentions of mathematicians.

There is some guidance concerning other introductory books on mathematics in the 'Further Reading' section at the end. However, special mention needs to be made here to the similarly named *An Invitation to Mathematics* by Norman Gowar, which has a related but different intention to this book's. Norman Gowar's book is a doorway into mathematics, with the door open, whereas this book is a mat with WELCOME on it placed in front of the doorway.

There are mny references in this book to events in the history of mathematics, so it is essential to realize that I am *not* an expert on such history. Whenever I have needed to refer to mathematical history I have consulted some of the books on the subject and chosen one of the differing accounts to be found there. Therefore this book may not be reporting history the way it actually happened, but only the way it ought to have happened. We can hope that when the Great Loop of Time brings the events back again, they will take place in the proper way. Meanwhile, it should be noted that no serious attempt has been made to evaluate the various personal contributions to the progress of mathematics.

Historical imprecisions are not the only untruths to be found in this book. It is imporant to remember that, because mathematical statements are proved, they are the truest statements that we know. Consequently, large numbers of untruths have been written into this book in order to dilute its truth down to a strength closer to the everyday level. However, these untruths are not lies. For example, the account of the real Arthur Cayley's visit to the imaginary land of Ruritania cannot deceive the reader, so it's not a lie. There are probably also unintentional untruths in the book, and I would be pleased to be told about any such errors that are detected. Of course, as I have the final responsibility for what appears, those who have assisted me with this book can in no way be blamed for any of its inadequacies.

First among those whose assistance I wish to acknowledge is my wife, Ann. She has suffered recitals of the first, tentative versions and offered many useful suggestions for literary improvements and yet, somehow, she has consistently given her enthusiastic support for the project. My colleague Reginald Allenby then read the ameliorated draft and I must thank him for his criticisms of the mathematics and many valuable suggestions for further improving the text. Another important contribution to the book was made by the University of Leeds, which granted me five months of study leave that enabled me to write most of the book.

I also wish to thank many people who made useful suggestions which have contributed to the book. Among those who deserve my thanks are (in random order): David Salinger, Trevor Hawkes, Austin Bowers, Stanley Bashkin, Diogenes O'Rell, Janet Allenby, Joanna Dales, Frank Felsenstein, Grant Woods, Christopher Robson, T. G. Cowling, Margaret Bowers, Desmond Tetlow, George Gratzer, Janet Jagger, David Fairer, Wynford Evans, Douglas Jefferson, Brian Duffy, James Craggs, Christopher Lance, Brian Meek, Ronald Maude and John Boyling.

Finally, I thank the staff of Basil Blackwell who have helped and encouraged me with this book. Especially, I wish to thank Kim Picken, to whom the idea for this book belongs, and who made many useful suggestions about its content and enthusiastically encouraged its development, and Stephan Chambers for his encouragement and help towards its completion.

1

Mathematics – the Subject with a Difference: The Nature of Mathematics

J'ai seul le clef de cette parade sauvage.

Arthur Rimbaud, refrain of *Les Illuminations*

Do you want to know the truth? Politicians use rhetoric to try to persuade their audience to accept their message (whether it is true or not), theologians convey their beliefs and religious experiences, novelists write stories which communicate their understanding of society, critics offer their interpretations of works of art and relate them to masterpieces of the past, craft workers explain their practical methods and offer useful advice, journalists report and comment on what they have been told, lawyers present the evidence that supports their cases, scientists publish their observations and their deductions based on them, but only mathematicians write what they have *proved*. Therefore, although much of what is written and said is true, only mathematicians know that what they write is true because they can prove it. So, it may be claimed, mathematics is truer and better known than other forms of knowledge. However, this advantage is considerably offset by the fact that mathematics has a more limited cultural value than some other forms of knowledge. For example, theologians convey beliefs and experiences, but these refer to the ultimate destiny of humanity and the way that human life should be lived. On the other hand, mathematicians prove their assertions, but these assertions refer only to mathematics.

But what is a mathematic? Two mathematics, arithmetic and geometry, have existed since the time of Pythagoras (about 540 BC) and this plurality was represented in Euclid's celebrated *Elements* (about 300 BC), which contains three books of arithmetic as well as the familiar geometry. But written mathematics is a great deal older than these dates suggest; by about 1500 BC, the Babylonians already knew a large number of sets of numbers which are the lengths of the sides of right-angled triangles, and it is clear from the size of these numbers that they were not found by trial and error but obtained by a mathematical procedure. Because the right angle is the right angle for two walls to meet or for a wall to meet the ground, it seems probable that the Babylonians combined arithmetic, geometry and applications to building in their concept of mathematics. Indeed, it seems that mathematics developed initially as a study of measurement, with the measurement of area engendering geometry and problems of computation engendering arithmetic. This agrees with the dictionary definition of mathematics as 'the abstract science of space, number and quantity', and the dictionary derives the plural appearance of the word from the French *mathématiques*. This word is derived in turn from the Latin *mathematica*, though the Romans did not agree whether this was a feminine singular or a neuter plural. The uncertainty was due to the word's being borrowed from the Greek *mathematika*, which is related to the Greek *mathema*, meaning 'understanding'. In contrast, the word 'science' is derived from the Latin *scire*, which means 'to know'.

In fact, written mathematics seems to have originated in Egypt in the period 3000–1600 BC, but the archaeological evidence for this is more architectural than documentary, and mathematics continued as a practical combination of geometry and arithmetic throughout the Babylonian period (say, 1700–300 BC). The Babylonians also applied their mathematics to astronomical problems. In classical Greece mathematics was presented in a purely deductive form, designed for use by philosophers at elegant dinner parties rather than men on building sites wearing the Athenian equivalent of wellington boots. However, this trend was reversed before 212 BC by the first of the trio of truly great mathematicians, Archimedes. In addition to making calculations in geometry which foreshadow calculus, he studied the forces acting on levers and in water, thus initiating the topics of statics and hydrostatics in mechanics. At about this time, written mathematics had further independent origins in China and India, where advances were made in arithmetic (including the introduction of 'Arabic numerals', which are, in fact, Indian), and, to some extent, algebra. In the Roman Empire the Greek tradition of mathematics continued, and the geographer Ptolemy (about 150 AD) added trigonometry to the range of mathematical subjects.

The collapse of the Roman Empire halted mathematical work in Europe, but Arabian mathematicians collected mathematical results from Rome, India and China and used them as the starting point of their own investigations. in particular, the Arabian mathematicians developed algebra after about 750 AD, though the subject had a history that went back to the Babylonians, who had arithmetical techniques for solving some quadratic problems. Contact with the Arabians revived European mathematics after about 1200 and, by about 1550, the subject had gained another branch when Cardano's book *On Dicing* provided the disreputable first steps in probability theory, the basis of statistics. The seventeenth century was notable for the introduction of new branches of mathematics which broke away from the earlier traditions. First, in 1614, John Napier published his discovery of logarithms, which not only transformed practical arithmetic but also initiated mathematical analysis (the study of infinite processes) by his use of infinite series. In 1637, René Descartes introduced coordinate axes into geometry, which allowed algebra to be used in geometrical problems, thus creating algebraic geometry and paving the way for the remarkable geometrical innovations of the nineteenth century. These two significant advances in mathematics were then eclipsed by the publication (by about 1680) of the work of the second of the mathematical giants, Isaac Newton (1642–1727). In attempting to define speed and to obtain sums involving continuously varying quantities, he discovered calculus (as did Gottfried Leibniz at about the same time), which he then applied to problems of moving bodies, thus adding dynamics to the subject of mechanics. He also initiated physics, both practical and theoretical, which has been the source of influential mathematical problems ever since.

For the next century mathematics raced like a mountain stream down a gorge, but its course was broadened by Léonard Euler (1707–83), who was probably the most prolific mathematician of all time. In 1736 he discovered factors which limited the choice of afternoon walk over the seven bridges at Königsberg, and thereby he created graph theory, a subject that has gained in importance due to its relevance to computing. In 1750 he initiated complex analysis, the study of functions of a variable which can take the value $\sqrt{-1}$ as well as real numbers, a topic that arose from the difficulties experienced by medieval algebraists but which had a unifying affect on mathematics in the nineteenth century due to the work of Augustin Cauchy (1789–1857). In 1801 Karl Friedrich Gauss (1777–1855) signalled his arrival as the third, and possibly greatest, mathematical giant with the publication of his *Disquisitiones Arithmeticae*, which introduced many new and more deeply searching methods into arithmetic. In his lifetime Gauss dominated every branch of mathematics, and he created one new one, differential geometry, by applying calculus to geometrical problems.

The year 1850 brought two new triumphs, and with them two new branches of mathematics, when Arthur Cayley (1821–95) collected several topics under the heading of linear algebra by his work on matrices and George Boole (1815–64) discovered the mathematics of logic, which had long been sought by both mathematicians and philosophers. Boole's work was complemented by Georg Cantor's exposition of set theory in 1880, which included an investigation of the properties of infinite numbers. This work was the starting

4

point for the somewhat paradoxical discussion of the foundations of mathematics which has taken place during this century. At about the same time, Karl Weierstrass (1815–97) set the modern standard of rigorous proof in analysis and Henri Poincaré (1854–1912) (with others) created topology, which has been the most fertile branch of mathematics in this century. Since 1900 a large number of branches of mathematics have come into existence, but most of them are combinations of existing branches, such as linear analysis, in which the methods of analysis and linear algebra are cunningly combined for application to suitable problems. However, John von Neumann, who was one of the mathematicians who contributed to the invention of the computer, instigated a completely new branch, the theory of games, in 1947. This branch of mathemtics attempts to devise winning methods at intellectual games (such as poker) and it has led to new departures in mathematical economics.

There are yet further topics in mathematics which have grown from imperceptible origins over a long period, and two of these have become prominent recently. Numerical analysis, essentially the arithmetic of approximations, is vital for the proper use of computers, and the ability of computers to carry out the necessary calculations is probably also a reason for the revival of interest in combinatorics, which had a previous flowering in the early nineteenth century. Combinatorics, which has applications to probability, is concerned with counting the members of a large set by means of some smaller sets. The nature of the subject can be understood better by attempting to solve the following probem, which is of outstanding importance for all readers who own circuses.

Problem

In how many ways can one arrange a parade of three bull elephants and five cow elephants round a circus ring so that each elephant holds the tail of the elephant in front with its trunk but no bull holds the tail of another bull?

(Each chapter of this book contains one or two problems

and their solutions are all to be found at the end of their chapters.)

However, mathematics should not be thought of as a collection of topics to be studied in isolation. For example, it is impossible to study mechanics without calculus, and especially that part of calculus which treats differential equations, that is, equations involving two or more variables and their differential coefficients. But this relationship is to the mutual benefit of the subjects because, in return, mechanics supplies calculus with significant problems. Even practical arithmetic is not independent of other mathematical subjects because it has employed analysis for the calculation of logarithm tables, geometry for the construction of slide rules, and logic and electronic engineering for the construction of calculators and computers. Further, practical arithmetic leads to 'higher arithmetic', which is more commonly called 'number theory', and there the intention is to prove results about the integers. Some of these results require the use of analysis for their proofs, and others even need analysis in their formulations. Consequently, it is impossible to study number theory extensively without also studying analysis. This relationship also holds between number theory and algebra. In addition to its internal connections, mathematics also provides the theoretical parts of other subjects, such as physics, engineering, biology and geography. Topics like thee are always in some danger of slipping through the cracks between two departments. For example, a distinguished theoretical engineer was overheard saying sadly 'The mathematicians say I am an engineer and the engineers say I am a mathematician. And, what is more, both remarks are meant pejoratively.' There is even one subject, logic, in which mathematics overlaps with both philosophy and computer science, especially in the study of whether computations are possible.

The fact that there is an overlap between mathematics and computer science misleads many people into thinking that, now computers exist, mathematics has been replaced by computer science and that no mathematician ever works without a computer, For my part, I have recently upgraded my computing system, by adding a very basic pocket calculator to my set of logarithm tables. They are both used very sparingly. However, for any mathematician who needs to perform long or complicated calculations, access to a computer is essential, and so is knowledge of a suitable computing language. But there is no immediate danger of computer science replacing mathematics because no computer is capable of solving a scientific or industrial problem – indeed, not many of them can even make a decent cup of tea. The problem needs first to be transformed into a mathematical problem – this is called *modelling* – then the problem needs to be put in a form which is suitable for computation, and a computer program devised to perform the calculation. Only then can the computer take over the problem and give a numerical solution, with the speed and accuracy which is typical of a computer, but which would be utterly astonishing in any other context. This account illustrates the fact that computing science is not supplanting mathematics but is generating new mathematical problems and offering new resources for their solution. However, although it has mathematical parts, computer science is not itself a part of mathematics. Because it is concerned with the design and use of electronic computers, it is really more closely related to electronic engineering.

Unlike computer science, statistics (the science of deduction from numerical data) is a branch of mathematics, but it will not be considered here as it deservedly has an *Invitation* of its own. Applied mathematics will be another significant omission from our discussion, though only partial. A difficulty with the subject is that it has no agreed definition, If applied mathematics is the collection of applications of mathematics, then the mathematics involved is mostly pure mathematics and statistics, because items of 'applied mathematics' are either exactly the mathematics the problem requires or just about useless, whereas the non-applied mathemtics is more generally available for appliction. Furthermore, as mathematics is used in almost all scientific work to test hypotheses by means of predictions

7

which can be the subject of experiments, applied mathematics would include most of science by the working of this definition. The experiments are, of course, subject to the **universal scourge called** *Sod's Law* or, more euphoniously,

The Law of Depravity   Every entity tends to act in order to falsify any prediction made about it.

Alternatively, applied mathematics is all the mathematics that has been applied. At present, this covers more than half of mathematics; the remainder will probably be used in the future. In *Invitation to Engineering*, Eric Laithwaite devotes chapter 5 to mathematics and uses it to point out that engineers need more than elementary mathematics, irrespective of the computing power available. David Whynes, in *Invitation to Economics*, recommends that the amount of mathematics that an economist needs is 'The more, the better'. However, he also warns of the danger of becoming too interested in the pure mathematics and never returning to the economics. A similar, and possibly greater, danger is that (say) an economist develops computing skill for some application and becomes so fascinated with the power of the computer that he never emerges from the computing laboratory. (It should be clearly understood that 'he' and 'she' in this book are of common gender, and that 'he' often turns out to be 'she', and 'she' to be 'he', when some particular person is referred to.) The computer is now capable of so many statistical procedures that it would be possible to analyse the numbers of a page of the telephone directory until the last trump is played. Presumably, the reading of telephone directories would then become part of applied mathematics if it is to be regarded as the collection of all the parts of other subjects which are amenable to mathematical treatment. With this definition, applied mathematics contains theoretical physics, mathematical biology and mathematical sociology, but there are necessarily some wide divergences between the theoretical treatments of these subjects and observed results, in accordance with the following rule.

The Second Law of Depravity   The predictability of an entity is inversely proportional to the square of its quantity of life.

There seems to be only one topic which is universally regarded as applied mathematics: theoretical mechanics, a topic which will be considered in chapter 3. A difficulty in studying theoretical mechanics is that very few shops stock theoretical spanners sized in the SI units which now prevail. A final complication in the determination of applied mathematics is the existence of topics in pure mathemtics, like the study of differential equations, which are of greater interest to those who are applying mathematics. In fact, these topics are sometimes taught as pure mathematics and sometimes as applied mathematics, according to what is convenient at the time.

As a general rule, apart from the intrinsic value of an application, mathematics is best studied as pure mathematics, with the material classified according to the kind of method that is used, wheres applied mathematics is better at revealing the reason for studying the item concerned. This kind of motivation is most important for subsidiary courses designed for students of a subject other than mathematics, but such subsidiary courses usually have a barely adequate allowance of time, leaving no time for any motivating reason. It seems less sensible to design specialist mathematics courses to emphasize such motivating reasons, because anyone who needs reasons based on (say) physics for studying a part of mathematics would seem well advised to study physics instead. However, some would-be physics students are diverted to mathematics because most physics courses are primarily concerned with practical physics, and so they tend to exclude the student who cannot cope with the practical work because of having two left hands (or who is left-handed and has two right hands). Perhaps that is why the shortage of physicists is even worse than the shortage of mathematicians!

To be more exact, there is an acute shortage of physicists, but a superfluity of mathematicians, provided that anybody who can piece together a mathematical proof is regarded as a mathematician. For example, we assume immediately that almost anybody can work out the cost of carpeting a room provided that they are told the room's measurements, the cost of the carpet material and the fee for laying it. The calculation of this cost is really a simple mathematical proof in which the cost is deduced by mathematical reasoning from the premises. In the sixteenth century, only an educated minority would have been capable of this deduction, but, by comparison, we are all mathematicians now! Consequently, we now divide mathematics into the *elementary mathematics* of domestic and commercial life, and other trivial pursuits, and *(higher) mathematics*, to which this book extends an invitation. In general parlance, only those who study higher mathematics are described by the word *mathematicians* and regarded as being, in some way, members of an elite class. We shall discuss whether this class really is exclusive in chapter 10, where we review various controversial aspects of the lives of mathematicians. But there is no controversy concerning one statement: there is a severe shortage of mathematicians, in this more restricted sense.

So that is mathematics. On the subject of cheeses, Hilaire Belloc wrote: 'Were I writing algebra (I wish I were) I could have analysed my thoughts by the use of square brackets, round brackets, twiddly brackets, and the rest, all properly set out in order so that a Common Fool could follow them.' In this passage Belloc makes it clear that in algebra, unlike in English, sentences must be constructed so that the symbols are combined according to predetermined rules, that is, its grammar must be unambiguous. But that is not enough. An old history examination paper asks: 'Did Julius Caesar pribble his enemy?'. The grammar of this question is perfect and its subject is well known, but we cannot answer the question at all unless there is a meaning for 'pribble'. If we do find 'pribble' in a dictionary we still cannot answer the question because Julius Caesar had many enemies, and he even had some sharp disagreements with his friends. So which enemy was intended? No statement can be true, or (for that matter) false, unless it is meaningful and unambiguous. It is very easy to write a meaningful and unambiguous mathematical sentence (and even easier to write a meaningless, ambiguous one), but it is very hard to write a meaningful and umambiguous sentence in English. Consider, for example, the simple assertion 'All lions are brown.' Does it refer to all members of the species *Felis leo* or does it ignore the lionesses and cubs? And what about the red lion outside the hostelry? Is that to be considered, and, if so, is it actually brown?

On the other hand, English (or any other natural language) is much more suitable for writing poetry. Admittedly, Shakespeare started to write a soliloquy in mathematical language:

$$(2b) \lor (\sim (2b)) \equiv ?$$

but even he experienced considerable difficulty and expressed the rest of Hamlet's thoughts in English. The reason that mathematical language is unsuitable for poetry is that it is too precise, as in the formula

$$F = C(1+r)^n$$

which determines the final sum F units obtained by investing a capital sum of C units for n years with compound interest at rate r. (Technically, we are contrasting the denotative merits of mathematical language with the connotative merits of Engish.) The precision of mathematics is the reason why it is the language of science. It records the result of an experiment precisely, then predicts with equal precision the result of a future experiment precisely, then predicts with equal precision the result of a future experiment, subject to the validity of the background assumptions and the accuracy with which the mathematics actually corresponds to the result of the experiment.

Mathematics is also its own language, however, and therefore the language of mathematics progresses as the subject does, but in a different manner from English. The structure and meaning of English sentences have changed as the centuries have passed in order to confuse the ordinary reader and to provide scholars with the pleasant task of elucidation. For example, when King James II said that St

Paul's Cathedral was 'amusing, awful and artificial', Sir Christopher Wren was, no doubt, delighted that this work was regarded as pleasing, awe-inspiring and skilfully executed. In the unlikely event of Queen Elizabeth II using the same words about his masterpiece, Sir Christopher would have to look for consolation in his lasting reputation for his mathematics, such as his work on the cycloid. Mathematical language changes whenever a new result is proved, by combining the former meanings. For example, when it is proved, for any real variable x and any real number c, that

$$(x + c)^2 = x^2 + 2cx + c^2,$$

the results which are already known to be true for $(x + c)^2$ are afterwards known to be true for $x^2 + 2cx + c^2$, and vice versa. (It can reasonably be argued that both had these meanings all along, but we did not know that they had them until we proved the result.) The only possible change in existing terms in mathematics is to gain further meanings, but entirely new terms can be added to the language as well. For example, when it is shown that $\sqrt{-1}$ cannot be a real number but that the complex numbers hve an objective existence, then the term $i = \sqrt{-1}$ is gained.

Mathematics as a language, however,, must be carefully distinguished from mathematical notation and terminology. For example, the symbols 2 and II both represent the number two, and we can also describe a set of two elements as a pair or as a couple. Even worse problems of terminology arise if we read mathematics from a text in a foreign language. However, it is highly inconvenient (though not impossible) to write mathematical results without using English (or some other natural language) between the formulae and writing mathematical definitions in a natural language, though the ambiguity of natural language makes it difficult to formulate the mathematics properly.

The fact that each result in mathematics enriches the language has the effect that mathematics needs to be learned (and taught) in something resembling logical order. For example, a result about the square root of the polynomial $x^2 + cx + c^2$ will need to come after a proof that $x^2 + 2cx + c^2$

$= (x + c)^2$. This does not mean that there is a unique order in which the results must be learned, because, for example, most elementary results in algebra are entirely independent of results in elementary geometry. The order in which mathematical items are learned is usually not the order in which they were discovered, because result are usually glimpsed through a thicket of extra conditions, then generalized with more complicated proofs, and simple proofs that go to the hearts of the results are only discovered later.

On the other hand, the historical development sometimes suggests an easy way to approach a topic which contains some notable difficulty. For example, the strict definition of the set of complex numbers is too sophisticated to teach to those who are just beginning to learn elementary algebra, even though the solution of some quadratic equations requires the use of complex numbers. Historically, algebraists used the complex numbers in this context, though without any confidence in what they were doing, by assuming that they behaved broadly like the real numbers, though not in all details, because the use of the rules concerning 'greater than' readily leads to contradictions when they are applied to complex numbers. Obviously, this approach is unsatisfactory, but it suggest that complex numbers can first be used with the aid of a list of properties that the numbers satisfy, and this list of properties can later be verified when the strict definition is studied. In the meantime, the list of properties can be used to *prove* results like the following:

if x is a complex variable and c is a complex number, then
$$(x + c)^2 = x^2 + 2cx + c^2.$$

However, no matter what background assumptions we make, we shall never be able to prove that 'The cat sat on the mat.' First, we shall have difficulties with the terms of the sentence, such as the difficulty of distinguishing between a cat and a tiger cub. Second, although we *saw* the cat, *smelt* its feline scent, *stroked* its fur and *heard* it purring, we can only claim to have *observed* it, not to have *proved* that it was there. We could discuss the alternative proposition 'We

13

observed that the cat sat on the mat', but we would only be adding to our philosophical difficulties, not providing a *proof*.

The difference between an observation and a proof assumes considerable importance if we wish to pass on a discovery to a friend. If it is a mathematical discovery which we have proved, we can start with the most basic ideas and agree with our friend about the names *one*, *two*, *three*, etc. and the symbols 1, 2, 3, etc. After this we can agree on various terms related to logic and set theory such as *equals* (=), *not* (∼), *belongs to* (ε), etc. and then build up agreed terms and symbols for all the mathematics needed for our proof. Then, at last, we can tell our friend the new proof and he will necessarily know as much about this discovery as we do. However, if we see the cat sitting on the mat, we can only report the fact to our friend, who will not *know* that the cat was there, but only *believe* that we had seen it. As we cannot prove, even to ourselves, that the cat was present, we have no proof to pass on to somebody else which will convey our knowledge to them.

'How can I convince somebody that the cat is there?', I asked myself one afternoon, and I looked at the cat's favourite place by the radiator. My observation then told me that I would never convince anybody about it, because the cat had departed in search of alternative entertainment, and his place had been taken by a very small man. In response to my astonished expression, the tiny man introduced himself as Zxt and explained that he had just arrived from the galaxy Knonot. After saying that, he vanished, but he reappeared at a more normal size on the other side of the room and sat down in the armchair. After introducing myself, I asked him how he crossed the room, and he explained that, to avoid the furniture, he stepped back into a parallel three-dimensional space and then returned to ours. I then realized that I was not seeing all of Zxt, but only his intersection with our three-dimensional space. I complimented him on his command of English and he explained that he had learned it from the radio. While he was saying this, he looked round my study and he then asked me about the saucer of milk that had been put down for the cat. I attempted to explain what it was and the purpose it served, but I failed utterly with my description of a cow, and my attempt led us to a discussion of geometrical figures. By simple demonstrations and definitions, we were readily able to agree on the identity of triangles, cuboids, circles and spheres, but I could make no sense of his account of the hypersphere (the set of all points equidistant from a given point in four-dimensional space).

I quickly switched the discussion to arithmetic and, once we had established the names and symbols for the integers, we easily agreed about the basic operations for the integers (that is, the positive and negative whole numbers), the multiplication tables, the definition of vulgar fractions and the arithmetical rules for vulgar fractions. So that we could later discuss calculus, I decided that I would try to introduce the irrational numbers, for which the uniqueness of the decimal representation would be helpful. As this is a consequence of the Division Algorithm, I then proposed the following statement.

The Division Algorithm If a is an integer and b is a positive integer, then there exist unique integers q and r such that
$$a = bq + r \text{ and } 0 \leq r < b.$$

In other words, for two non-zero integers, there is a unique quotient q and a unique remainder r.

We seemed to agree about the existence of q and r, but when I argued that the assumption that a also had the integer quotient p on division by b with integer remainder s with $p \neq q$ led to the statements that $|s - r|$ (that is, the positive value of $s - r$) was both less than b and not less than b, Zxt expressed puzzlement. I explained that the pair of statements contradicted each other, but Zxt objected that there was no contradiction, indeed they were the same statement. it was my turn to be puzzled, and Zxt offered the explained that, in a radio programme, he had heard the mathematical statement

that function
Is smother'd in surmise, and nothing is
But what is not.

15

He had obviously gleaned this item of non-existential philosophy from a certain Scottish play, so to convince him that this is an aberrant use of the word 'not', I appealed to the dictionary and found the following.

> *not*   a negative adverb,
> *negative*   expresses denial,
> *denial*   a declaration that the statement is not true.

I started to explain that 'not' was such that 'not not true'meant 'true', and this led Zxt to ask if 'not' meant the same as 'upside-down'. While I sought to elucidate further, Zxt was reminded of another engagement and, promising not to return later, he vanished.

I feel sorry for Zxt. He seems such a promising mathematician but, unless he transcends the logic which the present inhabitants of Knonot understand, he will never be able to learn about calculus and the other parts of 'higher' mathematics. The reason for this is that the meaning of the (otherwise undefinable) word 'not' is essential for the use of ordinary logic, which is called *Artistotelian logic* after the philosopher Aristotle, who formulated its basic laws in about 340 BC. The fundamental assumptions of Aristotelian logic are the following pair.

The Law of the Excluded Middle   Any meaningful statement
    is either true or not true.

The Law of Contradiction   Any meaningful statement cannot
    be both true and not true.

Because Zxt and the other Knonoters do not understand these two laws, they cannot prove results in mathematics which we have already proved. But are we certain that we know everything that we need about logic in order to prove all possible results in mathematics? In fact, Aristotelian logic has been somewhat refined by medieval logicians and then by nineteenth-century logicians such as Augustus De Morgan and George Boole, and the foundations of mathematics have been precariously erected on a theory of sets very carefully developed from Georg Cantor's initial idea. Our question then becomes: 'Can all true mathematical propositions be proved from our present knowledge of logic and set theory?'. This question has a subtle answer, which we shall approach gently by first considering the forms that mathematical results take.

The familiar forms of results include geometrical and algebraic identities, formulae in calculus, inequalities and solutions of various equations. These can be described as positive results, but there are also negative results, which assert that something is impossible. For example, there are no real number solutions of the equation $x^2 + 1 = 0$. We can express this as a solution of the problem by saying that the set of real solutions of $x^2 + 1 = 0$ is the empty set, $\emptyset$. However, many old mathematical problems have not been solved yet, and it is tempting to suppose that they can all be solved if enough time is spent on them. This temptation is one of the many which must be resisted, because it was proved in the early nineteenth century that there are quintic (fifth degree) equations that cannot be solved by means of the usual arithmetical operations and the extraction of roots, such as fifth roots. This was the first of many theorems of the type which determines whether some problem has a solution, and what form the solution has. These results sound purely theoretical, but they actually have great practical value because they indiate whether and how an approximate solution to the probem can be found with a computer. As a consequence, we could be reasonably satisfied with the horizons of mathematics if we could be sure we could settle all questions about the existence of solutions even when we could not actually obtain them.

The truth, however, is more complicated and less satisfactory than that. In 1930, Kurt Gödel proved that there are statements about arithmetic which can be neither proved nor disproved from a set of axioms for arithmetic. As arithmetic permeates all mathematics, this means that in any branch of mathematics there are questions we cannot answer at all; that is, they are *undecidable*. We might have hoped that no important proposition is undecidable, but, in 1963, Paul Cohen proved that two important propositions concerning the foundations of mathematics are undecidable. In other words, they are in some sense true, but we shall never be able to prove them. This implies that we cannot equate 'true' and 'provable', which brings us to the difficult question 'What is truth?'. The answer is clear enough, but needs to be phrased in accordance with one's personal philosophy: the *truth* is what God knows. But how can we *know* the truth? Only, it seems, if we *prove* the result can we know it in such a way tht we know that we know it. So how could two mathematicians have different answers to the same problem? Unless there is something fundamentally wrong with mathematics, there are three possibilities:

1 at least one of them has made a mistake,
2 the problem has at least two different solutions (such as 'Find a root of $x^3 - 6x^2 + 11x - 6 = 0$),
3 they have used different and incompatible systems of logic.

However, it is possible to give a proof in mathematics, whereas it is impossible to prove anything in non-scientific subjects because these are founded on verbal statements, and the words are only fleeting symbols for shadowy sense impressions. The initial statements in the empirical sciences are derived from experiments so, although the mathematical arguments may be correct, the conclusions in the empirical sciences are vitiated by the fallibility of the observations in the experiments. In other words, results in empirical science are exactly as true as the immediate deductions from the experiments. The only subjects in which valid proofs can be given are the logical sciences, in which the results are proved from axioms using some system of logic. The logical sciences consist of mathematics and the mathematical parts of empirical sciences, computing science and philosophy. However, the results in the logical sciences are derived from the initial assumptions of the subjects, and the truth of these cannot be known as statements about the real world without using some subject outside logic and set theory. But in spite of all the deficiencies in its foundations, mathematics contains the 'truest' statements that we are aware of, the ones which are *necessarily* true whenever the assumptions hold.

A view of mathematics which was formerly popular is that the subject is only used for teaching to others, though the wide range of applictions of mathematics is now better appreciated. However, the mistake is now often made that someone who is working rapidly through some mathematical calculations must be a mathematician. Possibly she is, but often a mathematician is to be seen working slowly and intermittently, with occasional bursts of activity when the working is screwed up and thrown away. In these circumstances, the mathematician is rarely applying a well-known technique to a conventional problem, but instead she is trying to do something new. What did I hear? 'Surely all mathematics is known?' This question displays a fine medieval view of the state of mathematical knowledge: everything is already known and the aim of scholarship is to classify knowledge in order to reveal the relations between the parts. Why is this view taken of mathematics but not of chemistry, physics or biology?

One reason is that there are frequent news items about the uses of new chemicals, the discovery of new and more mysterious elementary particles and further observations of the courting rituals of the duck-billed platypus, but the language aspect of mathematics prevents such news items about mathematical discoveries. A second reason is the abstract nature of mathematical discoveries and a third reason is that mathematics is taught in something approaching logical order, so that general mathematical education stops well short of the areas where new results are possible. The rest of this book should convince the reader that there are large tracts of mathematics that are new to him and provide hints about how new mathematics is discovered.

The journal *Mathematical Reviews* tries to review all the mathematical papers and books that are published and it publishes over 40 000 reviews every year. The relevance of the flood of new mathematics for students of mathematics is that there is a steadily increasing quantity of mathematics to learn (though the time has long passed when it was assumed that students would learn something about each branch of the subject, let alone every result). This might serve to discourage students from studying mathematics, particularly as school children already put mathematics below first place in the popularity table! For example, Anonymous has expressed his opinion of mathematics in the following words.

> Seven plagues had Egypt,
> We have only three,
> Algebra, Arithmetic
> And Plane Geometry.

The same poet, in his noble *Valediction*, writes:

> No more school, no more sticks,
> No more rotten Arithmetics,
> No more Latin, no more French,
> No more sitting on the old school bench.

Such enthusiastic farewells to the subjects which, allegedly, best predict success at university! As well as to French, which always imparts a certain Gallic *savoir-faire*. Perhaps the introduction of 'Modern Mathematics' (or 'New Mathematics'), which is neither modern nor new, will restore mathematics to the pre-eminence it enjoyed in Athens, where the citizens cracked geometrical theorems for each other on the way to the theatre for the latest comedy by Aristophanes.

The history of Modern Mathematics starts with the fall from the Golden Age Standard at the beginning of the twentieth century, when geometry was no longer taught directly from Euclid's *Elements*, but from textbooks which attempted to prove less and to explain more. The difficulties experienced by the average scholar reduced the number of proofs given in the geometry course, and this was balanced by a rise in the descriptive and trigonometrical content. The mathematical arguments in arithmetic were mainly calculations, and the logical aspects of elementary algebra were not usually stressed, so the elementary part of the school course contained only hints of the idea of proof. Meanwhile, back at the university, heavily computational courses based on calculus were making way for more conceptual courses based on mathematical analysis and abstract algebra. By about 1960, many mathematical students were falling short in the blind leap across the wide gap from school to university mathematics, and the object of the Modern Mathematics movement was to narrow the gap in both content and style. The movement has scored some notable successes. The early introduction of set theory provides a mathematical language which allows a proper emphasis to be given to precise definitions. The fuller treatment of linear algebra introduces the required component of proof in the elucidation of the unfamiliar context of matrices and, at the same time, provides a valuable technique for solving or simplifying problems. However, some of the other introductions have been much less successful.

Solution

But let us not forget the elephants and their parade round the circus ring. Elephants are perspicacious pachyderms, each with its own individuality, so we have here a problem involving three identifiable bulls and five identifiable cows. Let us (notionally) stop the parade and stand by one of the bulls (it does not make any difference which one is chosen, because all possible parades can be formed round each bull). There must be a cow in front of him (a choice of five) and a cow behind him (a choice of four, as it cannot be the one in front of him). There are therefore 20 possibilities for this part of the parade. Now list the possible arrangements according to the sex of the other elephants in the parade, which can now be regarded as a queue of two bulls and three cows, and perform similar calculations for each one. There are, in fact, 72 possibilities for this part of the parade, so the total number of possible parades that can be formed is therefore 1440. However, the similar problem concerning three black and five white mice in a circular track would have only six arrangements, as the mice would not be individually identifiable. On the other hand, in a real circus ring with real sawdust, the number of possible parades might be much less than 1440. For example, one elephant might have a tail which is too sore to touch, then the parade would have to be cancelled. For our calculation the elephants must be sufficiently individual to be identifiable, but not so individual as to have idiosyncrasies. In fact, the calculation only applies to perfect *mathematical* elephants. This tale has a moral: study mathematics and you will be able to *prove* something, about nothing; study all other subjects and you will be able to *prove* nothing, about everything else.

2
Begin Here: Axioms

'Give me a place to stand and I will move the earth.'

Archimedes

Pythagoras was one of humanity's greatest thinkers. This fact is partially obscured from us by the advances in knowledge and understanding since his lifetime (he lived from about 569 to 500 BC) and by the collapse of his philosophical system late in his life. Any account of Pythagoras' thoughts has to be partly conjectural because he became the leader of a brotherhood of scholars who kept their findings secret, and consequently the exact authorship of any item among their discoveries is not known.

Pythagoras had travelled in many lands east of the Mediterranean to collect their accumulated wisdom. As well as various mathematical and astronomical items, he learned about magic rituals for obtaining knowledge, and also about mystical intuition – the process of gaining knowledge by long and profound thought. From Greece, nearer to his home in Italy, came the work of the earliest philosophers (Thales, Anaximander and Anaximenes) who attempted to apply reasoning to physical observations and who reached some very wild conclusions. The question of Pythagoras was 'What is the key to all knowledge?'. The random result of magic rituals inclined him to favour reasoning as the main method of generating knowledge, but intuition led him to believe that the simplest knowledge concerns the whole numbers (the integers), as we learned from Zxt in chapter 1. Pythagoras therefore proclaimed to his brothers that 'The integers are the key to all knowledge.' He was encouraged in this view by his discovery of the simple relationship between musical pitch and the length of the string sounding the note, and then he moved on to apply his ideas to geometry.

Here he made two important advances, as well as applications of his principle which it would be kinder not to mention. First, he discovered that he needed to make some assumptions to which he could apply logic in order to prove geometrical results and, second, he proved the theorem that bears his name and which is crucial in plane geometry. But it was also the seed of the tree which ruined the foundations of the School of Pythagoras.

Let us consider the length d units of a diagonal of a square of side 1 unit. Then, according to Pythagorean doctrine, d must be a rational number, that is, an integer or a vulgar fraction. (Vulgar fractions were acceptable because they were expressed in terms of integers.) Naturally, Pythagoras wished to know the value of d. If we apply the Theorem of Pythagoras to the right-angled triangle consisting of two sides of the square and the diagonal, we find that the diagonal is the hypoteneuse, so $d^2 = 1 + 1$. In other symbols, $d = \sqrt{2}$. Consequently $\sqrt{2}$ is a rational number a/b, where a and b are integers and b may be chosen to be positive, and we can assume that a/b is expressed in its lowest terms, that is, the greatest common divisor of a and b is 1. Therefore $a = b\sqrt{2}$ and, squaring, we deduce that $a^2 = 2b^2$. Because the right-hand side of the equation is even, the left hand side must also be even, so 2 divides a^2 and therefore 2 divides a. Therefore there is an integer c such that $a = 2c$. Then $2b^2 = (2c)^2 = 4c^2$, therefore $b^2 = 2c^2$. But, in this equation, the right-hand side is even, so the left-hand side is even, therefore 2 divides b^2 and thence 2 divides b. Therefore 2 divides both a and b, which is contrary to the assumption that a/b is expressed in its lowest terms. We conclude from this that $\sqrt{2}$, the length of a diagonal of a square of unit side, is irrational, that is, *not* a rational number. After spending a number of years trying to find a way of evading this conclusion, Pythagoras finally admitted that he had not found the key to all knowledge and, as knowledge is power, that the brotherhood would not gain all power after all.

The discovery of irrational numbers was also a setback to geometry, because the Greek system of numerals, which was more primitive than that of Roman numerals, was quite incapable of representing irrational numbers and there was no method of multiplying or dividing irrationals. however, Eudoxus, who lived from about 408 to 355 BC, provided methods for handling irrational numbers by approximating them by rational numbers, and these methods were incorporated rather tentatively in his textbook *Elements* by Euclid, who lived from about 330 to about 275 BC. Due to his textbook, Euclid has traditionally been mistaken for the leading Greek geometer of his period, but probably he should really be praised for a bold attempt to classify all the contemporary mathematics. The book certainly contained one important novelty: Euclid attempted to derive all the existing geometrical knowledge from a set of axioms (called 'postulates' in Sir Thomas Health's edition of the *Elements*). The axioms that Euclid used are the following.

1 A straight line may be drawn from any one point to any other point.
2 A terminated straight line may be produced to any length in a staight line.
3 A circle may be described from any centre at any distance from that centre.
4 All right angles are equal to one another.
5 If a straight line meets two straight lines, so as to make the two interior angles on the same side of it taken together less than two right angles, these straight lines, being continually produced, shall at length meet on that side on which are the angles which are less than two right angles.

Euclid intended his axioms to be obviously true, probably as physical observations, and he was well aware that the complicated Axiom 5 was not at all obvious.

William Wordsworth, in his autobiographical poem *The Prelude* (1805, published 1850), recounts a dream in which on a 'boundless plain/Of Sandy wilderness' he met an Arab mounted on a dromedary who carried

> underneath one arm
> A stone, and in the opposite hand a shell
> Of a surpassing brightness. . . .
>   the Arab told me that the stone . . .
> Was 'Euclid's Elements'; and 'This', said he,
> 'Is something of more worth' . . .
> An Ode, in passion uttered, which foretold
> Destruction to the children of the earth.

In accordance with 'Every man to his trade', it is no surprise that Wordsworth should give priority to the ode represented by the shell, but why, in this early reference to the idea of 'two cultures' (one based on english, the other on mathematics), does he pick on the *Elements* to represent science? The choice was almost certainly influenced by Immanuel Kant (1724–1804), who (with some earlier philosophers) asserted that geometry as expounded by Euclid was *necessarily* true. That is, the axioms were not merely true as statements in physics, but would have to be true in any universe whatever, and, naturally, the theorems of Euclidean geometry were therefore also necessarily true.

This categorical assertion prompted various mathematicians to investigate whether the unobvious Axiom 5 followed from the others. Euclid had been the first to try this, and he thought that he had proved a result which, together with Axiom 5, was equivalent to the following axiom, which was formulated by John Playfair (1748–1819).

Playfair's Axiom   Through a point not on a given line there
      is one and only one line parallel to the given line.

This seems to be an obvious result of physics, but it is not obviously a necessary truth. Further study of Euclid's axioms during the nineteenth century (with Playfair's Axiom replacing Axiom 5) made it clear that the *Elements* did not contain the perfect demonstrations envisaged by Kant. First, the basic definitions are faulty and attempt to define *all* the terms instead of basing the first and some subsequent definitions on undefined terms. For example, according to Euclid,

A *point* is that which has no parts, or which has no magnitude.
A *straight line* is that which ies evenly between its extreme points.

'Point' is not defined unless we already know the meaning of 'parts' or 'magnitude'. 'Straight line' acceptably requires a definition of 'points', but also needs the meaning of 'lies', 'evenly', 'between' and 'extreme'. None of these words is defined or listed as undefinable. Second, Euclid's axioms are insufficient for the result claimed. For example, many proofs use the intersection of two circles or of a line and a circle, but the axioms do not give any information about such intersections. Worse still, some result refer to a point between two others on a line, but the concept of 'between-ness' is not defined and no properties concerning between-ness are assumed. Finally, as we shall show later, the axioms are not a necessary truth, though this flaw in the *Elements* has been the starting point for some interesting mathematics, as well as the cause of some philosophical dismay.

The most serious affect of the disparity between the correctness of the *Elements* and its philosophical reputation has been the misery inflicted on many generations of school children, who have been forced to learn mathematical paragraphs which purported to be proofs but seemed irrelevant to the unhappy scholars, whose appreciation of a proper proof was correspondingly blunted. No wonder that some of the leading teachers of the late nineteenth century started the reform of geometry teaching! But can the imperfections in Euclid's *Elements* be corrected? There have been at least two attempts which seem to have succeeded and, of these, the version which David Hilbert (1862–1943) published in 1899 is closer in form to Euclid's original. Hilbert's underfined terms are *point*, *line*, *between* and

27

congruent and he listed his axioms in six sets:

1 axioms concerning connections between the basic terms, for example, two distinct points determine a straight line,
2 the properties of betweenness,
3 Playfair's Axiom,
4 the properties of congruences,
5 the Archimedian Postulate, which permits the measurement of line segments,
6 an axiom asserting that no further lines or points can be added to the plane without violating the other axioms.

Hilbert did not complete a proof that all these axioms are needed, but he showed that they can all hold at the same time by the use of the real number system (the set of all rational and irrational numbers).

Axiom systems have been usefully applied to geometry for over two millennia, so why is there no axiom system for arithmetic? There was a superstition, not entirely eradicated, that this is a consequence of geometry being a rigorous discipline while arithmetic and algebra are not. The confusion lies in the logic: arithmetic and algebra are felt by their detractors not to be rigorous because they do not possess axiom systems. So, has anyone ever tried to supply an axiom system for arithmetic? An interesting attempt was made in Britain by a lively and varied group of mathematicians: Charles Babbage, George Boole, D. F. Gregory, Sir John Herschel, Augustus De Morgan and George Peacock. Their attempt was a triumphant failure. No axiom system was found for arithmetic, but some of the positive results of their work were published in Peacock's *Treatise on Algebra* in 1830 and the ultimate result was the total transformation of algebra. The search for axioms for arithmetic had another aspect because other sets of numbers, containing the integers, were used systematically in the solution of some arithmetical problems. This idea had been used by Pierre Fermat, who lived from 1601 to 1665, and it was probably discovered by him.

Let us illustrate the idea by means of a problem of the

appropriate kind. For a given positive integer K let us find the non-negative integer solutions in both x and y of the equation $x^2 + 5y^2 = K$, if there are any. This is a trivial problem if regarded as algebra, with the solution $x = \pm\sqrt{K - 5y^2}$ for arbitrary y, but the problem is harder as an arithmetic problem requiring integer values for x and y.

We can study the case $K = 41$ by a method which can be applied to all the other cases. If we use the set of integers as our system of numbers, or 'domain' as it came to be called, then 41 is a prime and we cannot express $x^2 + 5y^2$ as a product of integers. However, if we consider the domain consisting of numbers of the form $a + b\sqrt{-5}$, where a and b are integers, then $x^2 + 5y^2$ can be factorized as $(x + y\sqrt{-5})$ $(x - y\sqrt{-5})$. In this domain 41 has the factorization $(6 - \sqrt{-5})(6 - \sqrt{-5})$, where the numbers $6 + \sqrt{-5}$ and $6 - \sqrt{-5}$ are irreducible, in the sense that they do not factorize further. Next we equate the two factors of the left-hand side to all the possible pairs of factors of 41, making use of the fact that the only units in our domain (that is, numbers by which every other can be divided in the domain) are 1 and -1. This gives the pair of equations

$$x + y\sqrt{-5} = A$$
$$x - y\sqrt{-5} = B$$

where $AB = 41$. By adding these two equations we obtain $2x = A + B$ and by subtracting the second from the first we obtain $2y\sqrt{-5} = A - B$. If the numbers A and B are unequal integers, then $y = (B - A)\sqrt{-5}/10$, which is a complex number. Consequently, the integer solutions of $x^2 + 5y^2 = 41$ come from the factorization as a product of complex numbers and, because $x = \frac{1}{2}(A + B)$ is required to be a non-negative integer, we also need $A + B$ to be an even, non-negative integer. The only factorizations of 41 which satisfy these conditions are $A = 6 + \sqrt{-5}$ with $B = 6 - \sqrt{-5}$ and $A = 6 - \sqrt{-5}$ with $B = 6 + \sqrt{-5}$. The first factorization of 41 gives $x = 6$ and $y = 1$ and the second gives $x = 6$ and $y = -1$, therefore the only non-negative integer solution of $x^2 + 5y^2 = 41$ is $x = 6$ and $y = 1$. Notice that the integer solutions of $x^2 + 5y^2 = K$, for any positive integer K, can be determined with the help of this working from the factorizations of K in the domain we have considered.

Problem

Find all the positive integer solutions of $x^2 - y^2 = 30$.

Returning to Peacock's treatise, we can now understand that the aim was to produce a set of axioms which could serve for the other domains of numbers but which could characterize the integers by means of some further axioms. It seems that some other applications of the axioms were contemplated as well, but at no time was it intended to *define* the integers or the real numbers. The object of the work was to give a list of properties that would characterize the integers inside the set of real numbers or the set of complex numbers.

Let us try to imagine the method which might have been used to find such a set of axioms. All the domains that were usually considered contained the integers, but none of them contained the set of rational numbers, so we shall not assume that the numbers in the set can be divided to give other numbers in the set. The properties of numbers under subtraction are difficult to write down, so let us evade the problem of subtraction for the moment. In modern terminology, let us consider a set of numbers D. If D is to be a domain according to the rough idea suggested above, we first ask that addition (written $a + b$) and multiplication (written $a \times b$ or $a \cdot b$ or ab) should be possible in D; that is, if a and b belong to D then so must $a + b$ and ab. Are there any obvious properties for addition and multiplication? First, it does not matter which way round either of these operations is performed; indeed, we can check our working for $a + b$ by also calculating $b + a$. This gives us two rules: for all a and b in D we have $a + b = b + a$ and $ab = ba$.

We can also check a sum like $232 + 21 + 492$ by calculating $492 + 21 + 232$. If we put in brackets to indicate the pairs that we add together first, we find that we really mean

$(232 + 21) + 492 = 232 + (21 + 492)$, an example of the *Associative Law*. The Associative Law is provable (from the definition of the integers) for numbers for both addition and multiplication, and the two laws have the consequence that no expression in numbers using addition alone or multiplication alone needs any brackets. So we add two more rules to our list: for all rules to our list: for all a, b and c in D we have $(a + b) + c = a + (b + c)$ and $(ab)c = a(bc)$.

In the set of integers and other domains it is possible to subtract, but this is more easily expressed in terms of adding negatives to obtain 0, which has the property that, for all a in D, the sum $a + 0 = a$. Now we can express the fact that subtraction is possible in d by the rule: for every a in D there exists a number $-a$ in D such that $a + (-a) = 0$. It follows from this rule and the Associative Law for addition that $-(-a) = a$. The rather awkward rules concerning subtraction in D now all follow from the rules for addition and the existence of negatives in D.

We want all the domains to contain the integers and we can ensure this by asking that the number 1 should belong to D, that is, we add the rule: D contains the number 1 with the property that, for all a in D, the product $a1 = a$. Now, $1 + 1 = 2$, $2 + 1 = 3$ and, by adding 1 to the last number obtained, we may obtain all the positive integers in this way. By the rule on negatives, if the integer n belongs to D, then so does $-n$ and, as D contains 0, the set D contains all the integers.

There is one more important property of the multiplication of numbers to be included; the product of two non-zero numbers is never zero; in symbols: if $a \neq 0$ and $b \neq 0$ belong to D then $ab \neq 0$. As it is important to be able to expand brackets, we also require the *Distributive Law*: for all a, b and c in D we have $a(b + c) = ab + ac$. In constructing proofs, the Distributive Law is the most important rule of them all, because it is the only one which connects the addition with the multiplication and, with the other rules listed above, it leads to all the standard rules for expanding brackets.

As the arithmetical operations have now all been con-

31

sidered, the natural next move is to draw up the rules for 'greater than'. These rules are easy to find, but it is also easy to show that the domains that contain complex numbers cannot have a relation which obeys the rules for 'greater than' which hold for the set of real numbers. This, therefore, appears the right moment to give a name to the set of rules which we have put together, but it is worthwhile to see if anything other than sets of numbers can obey the rules.

Let us start by considering the set P of all polynomials in a real variable x over the real numbers. P consists of 0, all the polynomials of degree 0, which are just real numbers, and all the polynomials of degree n for all the positive integers n. A polynomial of degree n is given by

$$f = a_n x^n + a_{n-1} x^{n-1} + \ldots + a_j x^j + \ldots + a_1 x + a_0$$

where $a_0, a_1, \ldots, a_j, \ldots, a_{n-1}, a_n$ are real numbers and $a_n \neq 0$. Note that the polynomial 0 does not have a degree and that the *degree* of a polynomial other than 0 is the greatest power of x with non-zero coefficient in the expression. Some careful calculations with polynomials soon reveal that P satisfies all the rules that we have set up with the polynomials 0 and 1 taking the places of the numbers 0 and 1. However, P is not a set of numbers, so the formulation of the rules above will not quite serve for P. To counteract this, let us write the definition of our domain to cover all cases like P, and let us choose a name to suggest the resemblance to the set of integers.

Definition Let D be a set on which is defined two (binary) operations called addition and multiplication. Then D is an *integral domain* if the following axioms hold for all a, b, and c in D (whether they are distinct or not):

$$a + b = b + a,$$
$$(a + b) + c = a + (b + c),$$

D has an element $0'$ such that, for all a in D, we have

$$a + 0' = a,$$

for every a in D there exists $-a$ in D such that $a + (-a) = 0'$,
$$ab = ba,$$
$$(ab)c = a(bc),$$
D has an element $1'$ such that, for all a in D we have $a1' = a$,
if $a \neq 0'$ and $b \neq 0'$ then $ab \neq 0'$,
$$a(b + c) = ab + ac.$$

(We write $0'$ and $1'$ to remind ourselves that these symbols act like 0 and 1, although they are not necessarily numbers.)

That the set of integers, the domains of numbers and the set of polynomials P are all integral domains would not be worthy of note unless it is possible to prove theorems about integral domains. Happily, that condition is satisfied. For example, it can be proved that for any integral domain D there exists an integral domain F which contains D in such a way that the addition and multiplication in F matches that in D but so that F has the additional property that division is possible by every non-zero element in F. The theorem actually says more than that, but the result indicates that it is possible to prove results from general axioms without knowing what the symbols actually *mean*. That is, the integral domains in Peacock's book are *abstract algebras*; the results about integral domains hold whenever the integral domain axioms hold, but in themselves they refer to no particular example. But we can add further axioms to the set of axioms for integral domains to construct a classification of integral domains which distinguish them by the kinds of theorem which hold for them rather than whether the particular integral domain is a set of numbers or polynomials or (as can happen) operations in calculus. Consequently, what started as a search for axioms for arithmetic has led to a system of axioms for a large part of algebra.

Solution

Now let us find the integer solutions of the equation $x^2 - y^2 = 30$. The method suggested by comparison with the equation $x^2 + 5y^2 = 41$ is to factorize 30 as AB and so obtain a

Begin Here

pair of equations

$$x + y = A$$
$$x - y = B.$$

From these two equations, $x = \frac{1}{2}(A + B)$ and $y = \frac{1}{2}(A - B)$, where $AB = 30$. The factorization of 30 into a product of prime numbers is $30 = 2 \times 3 \times 5$, from which it follows that one of A and B must be even and the other must be odd, therefore both $A + B$ and $A - B$ are odd. Consequently neither $\frac{1}{2}(A + B)$ nor $\frac{1}{2}(A - B)$ is an integer, therefore the equation $x^2 - y^2 = 30$ has no integer solutions. Notice that, had we found some solutions, it would have been easy to check by arithmetic that the pairs of integers were solutions, but the statement that the equation has no integer solutions requires a proof. Did you actually *prove* that there are no integer solutions of $x^2 - y^2 = 30$?

34

3
And Still it Moves: Mechanics

Nature and Nature's laws lay hid in night:
God said, 'Let Newton be!' and all was light.

Alexander Pope

In chapter 2 we discussed the choice of axioms for a non-physical mathematical theory, so now we investigate some of the axioms of theoretical physics. These form a significant part of mathematical physics because these axioms need to be confirmed experimentally and they must be chosen so that all known results in physics can be deduced from them. The further intention is to deduce new results from these axioms for future experimental investigation in order to check and refine the earlier work. This objective is a very ambitious undertaking, as, indeed, is the ultimate objective of any major subject. Because determining the axioms for the whole of theoretical physics might take us more than a few pages, let us restrict ourselves to just one aspect, the study of motion of objects ranging from the domestic to the astronomical, and so including all types of flying saucer.

The study of motion is essentially physics rather than mathematics if it is conducted informally with a chain of calculations occasionally checked by experiments, but it becomes the applied mathematical discipline of *theoretical mechanics* if it is studied as a formal series of deductions from a set of axioms, which are based on the results of earlier experiments. The mathematics used in the deductions should have a standard of rigour as close as possible to that of pure mathematics so that the results predicted give the best possible representation of the axioms. Just as the best pure mathematics is that which is applied, the best applied mathematics is pure, despite the contrary view expressed by a graffito seen in a physics laboratory:

When rigour is detected we know the subject is dead.

Perhaps the graffitist regards any topic in physics as dead unless it is in the exciting early experimental stage. But rigour, here in the form of the application of correct, logical argument, is as important in the growth of a theory as it is in a body. Young bones need to be supple to allow for rapid development and growth, but the bones of a mature body are rigid in order to support the flesh.

The mathematical subject of theoretical mechanics divides into topics according to the matter which is in motion. *Dynamics* investigates the motion of small objects and rigid bodies, and it entails a knowledge of *statics*, the study of conditions under which no motion takes place. The corresponding topics for liquids are *hydrodynamics* and *hydrostatics*. Moving gases are studied in *aerodynamics*, deformable bodies in *elaticity* and electrical forces in *magnetism*. All these topics are based on the axioms for dynamics, but have additional axioms of their own. Despite extensive rigour in their working, these topics are the subject of lively research, with many contemporary engineering applications. We shall consider only Newton's formulation of theoretical mechanics, that is, *Newtonian dynamics*. A competing theory is *special relativity theory*, which differs markedly from Newtonian dynamics for entities travelling near the speed of light, although at lower speeds the results of the two theories are approximately the same. Consequently, even near the speed of light, the results for Newtonian dynamics are useful in special relativity theory as a first approximation. In both Newtonian dynamics and special relativity theory, much use is made of numerical methods involving approximations, in contrast with most topics in pure mathematics. The undefined terms in Newtonian dynamics are *time*, *space*, *mass* and *matter*. The matter in a dynamical system forms a number of *bodies*, so dynamics concerns moving lifeless bodies, just like undertaking. For each of the undefined terms we need axioms to specify their basic properties and we also need methods and units for measuring them.

Axiom 1   Time is represented by a real variable.

This statement implies that the time is the same throughout space, independent of the presence of matter or the motion of bodies. Time is to be measured by an atomic clock – a digital watch is a crude specimen – and the SI unit of time is the *second* (think of SI as 'standard international'), which is a certain number of periods of oscillation of a caesium atom. In special relativity theory, clocks give different times if they are moving relative to one another, but the diffeence is negligible over a short time or if their relative speed is much less than that of light. Consequently, although some experiments favour the adoption of special relativity theory, Newtonian dynamics is suitable for almost all motions in the solar system of at least a domestic scale.

Axiom 2   Space is a three-dimensional Cartesian geometry
with respect to a fixed set of rectangular Cartesian axes.

A Cartesian geometry is a Euclidean geometry defined in terms of the set of coordinate axes according to the methods of René Descartes. In a three-dimensional geometry there are three mutually perpendicular axes and each point is represented by the unique traid of real numbers, the *coordinates*, which are the directed distances of the point from the planes through the axes. Here, 'directed' means that one direction is given a positive sign, the opposite direction a negative sign. Distance and angle are defined by formulae in the coordinates and all the methods of Euclidean geometry and trigonometry are available for measuring lengths and angles. The SI unit of distance is the *metre*, which is a certain number of wavelengths of radiation of krypton. The SI unit of angle is the *radian*, so that the radian measure of an angle APB at a point P is the length of arc of a unit circle with centre P between the lines PA and PB. But where are the fixed axes? According to hypothesis, there is a fixed space, the *ether*, containing the axes, and all moving bodies move relative to the ether. However, it seems to be much more difficult actually to find the ether than to find an invisible spook at midnight in a thick fog. Special relativity theory ignores the ether and considers only relative motion. However, the changing position of the stars is apparently so slow that the sun can be regarded as fixed for calculating motions which last for only a few hundred years, so Newtonian dynamics can be used for these motions.

In order to continue our discussion of dynamics we now need some definitions from *kinematics*, the geometry of motion. This topic is based only on the axioms about time and space, so it is better regarded as an interpretation of some pure mathematics than as a topic in applied mathematics. We shall avoid difficulties associated with three-dimensional space (such as the use of *vector analysis*) by limiting our equations to one-dimensional space, that is, to a line. But let us first consider an example. Near York, two cars joined the London to Edinburgh road at the same moment. The sports car drove at an average speed of 55 miles per hour (88 km per hour) and the economy car drove at an average speed of 30 miles per hour (48 km per hour). Which car arrived at Edinburgh first? Yes, of course, the answer is obvious. The economy car. The sports car went to London! Even in one dimension, a knowledge of the speed is not enough; the *direction* must also be known. The speed and direction taken together form the *velocity*.

Suppose that initially, when the time t is 0, a moving point P coincides with the point A. Let s be the directed distance of P from A at time t. Then the average velocity of P up to time t is the directed distance divided by the time taken, that is, s/t, which is measured in metres per second, denoted by ms^{-1} in SI units. However, the velocity of P at the time t can only be defined by using differential calculus, which we shall consider later. For the moment we shall introduce only the notation. The velocity v of the point P is given by the differential coefficient of s with respect to t, which is denoted by ds/dt or $\dot{s}$. However, unless the velocity is constant, we also need to know the rate of change of the velocity, which is the differential coefficient of the velocity v with respect to t, and it can be denoted by any of the symbols $\dot{v}$ or dv/dt or $\ddot{s}$ or d^2s/dt^2. Then $\dot{v}$ or $\ddot{s}$ is the *acceleration* a of the point P at the time t. The SI unit for acceleration is metres per second per second, denoted by ms^{-2}. When the acceleration a is a constant, the velocity v and directed distance s of the point P at time t are given by the following formulae, where u is the velocity of P at time $t = 0$.

$$v = u + at,$$
$$s = ut + \tfrac{1}{2}at^2,$$
$$v^2 = u^2 + 2as.$$

These formulae are only valid when the acceleration a is *constant* and it is wrong to use them in any other case. (More marks are lost in mechanics examinations for this mistake than for any other!) Also, these formulae apply only to motion in one line, and a change of velocity along one line is necessarily a change of speed. However, this is not the case for motion in a plane or in space. To illustrate this, let us consider the usual controls of a car. Obviously, the accelerator, the clutch and the gear lever are designed to alter the amount of the engine's motive force which is used, so they affect the acceleration of the car. Pressing the brake pedal reduces the car's speed, so it gives the car a negative acceleration, which is called a *retardation*. However, although rotating the steering wheel does not change the speed, it alters the direction of travel and therefore changes the car's accelertion. Clearly, the properties of acceleration are more subtle than those of velocity, so it is fortunate that Newtonian dynamics does not require a knowledge of the rate of change of the acceleration.

Think of the view of the ground from a jet airliner cruising at its usual 600 miles per hour or so (268 ms^{-1} *à la mode du* SI). The engines are humming happily to themselves, the sun shines lazily on the far-spread scene of fields, farms, hills and villages and the plane hangs, seemingly, motionless above them. Such a distant view impresses the mind as essentially static and idyllic, yet such tranquillity soon tires the mind and we look for other diversions. Good, the cabin staff have put up a table-tennis table in the aisle. Let me serve first from the tail end of the table. A good, slow service, well into the middle of the table. And there goes my best smash, streaking through the air at about 650 miles per hour (about 290 ms^{-1}). But how did you return it so easily? Has the altitude lifted your game? Love–one. I use the same service again. Why did it miss the table this time? But the engines sound different, they are slowing down. That means tht the plane is being retarded while the ball is not, and that makes judgement of the flight of the ball very difficult. Let us continue the game when we are on some solid, stationary ground.

Is it really stationary, though? Surely, it is on Spaceship Earth, which is hurtling round the sun at a speed of about 29 600 ms^{-1}! Furthermore, the earth is spinning round once a day, giving a surface speed which varies from about 464 ms^{-1} at the equator to nothing at the poles. And yet, the table tennis we have all played in the past was not affected by the direction in which the game was played. In other words, we have observed that absolute velocity does not affect the motion of domestic objects, so it cannot be detected by such motions, though accelerations can be so detected. Because the earth is not travelling in a straight line but goes round the sun, there must be an acceleration of the earth towards the sun, which should therefore affect our games of table tennis. In fact, the earth's orbit does imply such an acceleration on terrestial objects, but the acceleration has a greatest value of about 0·006 ms^{-2}, so the effect will be negligible for any dynamical system with accelerations of about 0·1 ms^{-2} or above and which run for only a short time.

As the kinematical equations above indicate, a small constant acceleration over a long period can have a large effect. Similarly, if a terrestial object were not subject to an acceleration towards the centre of the earth, it would fly off along a tangent to its line of latitude as the earth spun, providing a very cheap space launching programme. This acceleration has a greatest value of about 0.033 ms^{-2}, so dynamical systems with significant accelerations of the order of 0.1 ms^{-2} would need calculations referred to coordinate axes which were fixed in direction with the origin at the centre of the earth, or even fixed relative to the sun. Our general conclusion from these last considerations is that we may regard coordinate axes fixed with respect to the earth as approximately stationary only if we are considering a motion with large accelerations and which lasts for a short time.

We have not yet considered the properties of mass, so let us continue with the following axiom.

Axiom 3 The mass of a constant body of matter is independent of time, space and motion.

The SI unit for mass is the *kilogram*, denoted by kg, which is the mass of a lump of platinum and irridium which is preserved very carefully in Paris. The method of measurement of mass will be considered later. however, a body need not have a constant mass. For example, a rocket in flight is continuously burning a large quantity of fuel, so the amount of matter in the rocket is decreasing and therefore its mass is decresing, in accordance with Axiom 3. On the other hand, in special relativity theory, mass increases with speed (so, for example, joggers become *heavier*). As special relativity theory differs from Newtonian dynamics in its assumptions about time, space and mass, we shall in future consider dynamics only according to the hypotheses of Sir Isaac Newton. Also, we shall only consider the motion of *particles*, that is, masses concentrated at one point. This is much less restricting than it sounds because, in its motion about the sun, the earth is small enough to be regarded as a particle, and any large body can be regarded as a particle concentrated at the geometrically determined *centre of mass*, provided that, in this case, we are not interested in the positions of the other points in the body. For example, a high-board diver can be regarded as a particle in his motion through the air so long

41

as we only try to locate his centre of mass and ignore his twists and somersaults around it.

We now state the next axiom in terms of particles.

Axiom 4: The First Law of Motion   A particle moves with constant (perhaps zero) velocity unless it is acted on by a *force*.

This axiom defines force as the cause of change of motion, but neither defines nor quantities force. Towards this end we define the *momentum* of a particle as the product of its mass and its velocity. Because velocity is a vector (that is, it has a direction as well as a magnitude) this implies that the momentum is also a vector.

Axiom 5: The Second Law of Motion   The total force acting on a particle is proportional to the rate of change of its momentum, and the factor of proportionality varies only with the units used.

Before we look at the theoretical consequences of the Second Law of Motion, it is worth noting that it has the immediate consequence for a mechanical system on which there is no total force that the momentum is constant. it might be thought that the First Law of Motion implies that, if there is no total force, then the velocity is constant, but that axiom applies only to a single particle.

Problem

A girl found that a frozen pond on which she intended to play was so slippery that she should gain no grip on it in any way, so she decided against sliding on it. Instead, she climbed a tree which overhung the pond. Unfortunately, she climbed too far out on a branch and it broke. As a result, she found herself sitting in the middle of the pond holding a piece of the broken branch. How did she get off the pond without any assistance?

Because velocity is a vector, so is the momentum and its rate of change, that is, its differential coefficient with respect to time. Consequently, by the Second Law of Motion, the total force acting on a particle is a vector. The following axiom tells us how to find the total force acting on a particle in terms of the various forces which are acting on it.

Axiom 6   A force is a vector and the combined effect of two forces at a point is their vector sum.

This axiom implies that a force is determined by its components in three mutually perpendicular directions and that the sum of two forces is obtained by adding the corresponding components. From this last fact follow all the results of *statics*, such as conditons for a mechanical system to remain at rest and the laws of the lever, which were discovered by Archimedes. Let us investigate the meaning of Axiom 5 for a one-dimensional motion of a particle. Let m be the mass, v the velocity and H the momentum of a particle acted on by a total force F on a certain line. Then, by Axiom 5, there is a constant k such that $F = k(dH/dt)$.

Let us consider a particle which has a constant amount of matter and therefore a constant mass m, by Axiom 3. Because $H = mv$, the rules of calculus tell us $dH/dt = m(dv/dt)$ and, as we noted earlier that dv/dt is a notation for the acceleration, a, of the particle, therefore, in this case, $F = kma$. Because the constant k varies only with the units and the units of mass and acceleration are already chosen, we choose the unit of force such that $k = 1$ for the SI unit. This is the *Newton*, denoted by N, which is the force which accelerates a constant mass of 1 kg at a rate of 1 ms^{-2}. Therefore, if the force is measured in Newtons, the *equation of motion* is

$$F = \frac{dH}{dt}$$

or, if the amount of matter in the particle is constant,

$$F = ma.$$

As these equations can be regarded as a definition of total force, there is no need to look for experimental evidence to support Axiom 5. However, the next axiom is either a

43

brilliant deduction by Newton about the fall of an applie or the result of his careful study of existing experimental and observational evidence. The logical explanation seems more likely to be true than the romantic one, but all subsequent experimental evidence tends to confirm the truth of the axiom.

Axiom 7: The Law of Gravity   The gravitational attraction between two particles of masses m_1 and m_2 which are distance r apart is a force $\gamma m_1 m_2 / r^2$, where γ is a universal constant, determined only by the units used.

Experimental observation gives the value of γ as about $6 \cdot 67 \times 10^{-11}$ in SI units. A calculation using Euclidean geometry and calculus, in accordance with axiom 2, shows that a sphere of uniform matter has the same gravitational attraction as a particle of the same mass at the centre of the sphere. The earth is not quite a sphere as it has a radius at the poles of 6357 km and a radius at the equator of 6357 km but the same principle holds for the earth, though there is a negligible gravitational attraction towards the equator on the earth's surface.

The gravitational force of the earth on a particle P of mass m is called the *weight* of P. If P is h metres above the earth's surface at a place where the earth's radius is r, then, by Axiom 6, the weight of P is $\gamma M m (r + h)^{-2}$ Newtons, where M is the mass of the earth. If $g = \gamma M r^{-2}$ then g varies only with r, that is, g varies only with the latitude. For a relatively small value of h, the value of $(r + h)^{-2}$ is approximately r^{-2}, therefore the weight of P is mg. In this formula g is a *local* constant, varying with the latitude, and g represents the acceleration due to gravity of any particle near sea level at that latitude. The constant g has the value $9 \cdot 8321$ ms^{-2} at the North Pole, $9 \cdot 8119$ ms^{-2} at London and $9 \cdot 7810$ ms at the equator, so a given particle is heaviest at the North Pole and lightest at the equator. To make it lighter still, carry it up a mountain of height h at the equator, then the weight is somewhat less than its value at sea level, though, after carrying it up, it might seem much heavier. In other words, the gravitational constant g varies with the latitude and the height above sea level, but it is constant enough to be used in measuring mass.

Suppose that we wish to know the mass m of an object A and we have a 1 kg weight K. Then we balance a steel lever on its mid-point B, hang K one unit of length from B on one side, hang A on the other side of B on the lever and adjust the position of a until the system is at rest. (Theoretically, the lever should be placed east to west to avoid the attraction towards the equator.) Now we measure the distance AB as d units by geometrical means, and we deduce from the results on statics that the weights of A and K have the same moment about B, that is, that $1 \times g = d \times mg$, therefore $m = d^{-1}$ kg.

Now that we know how to make all the basic measurements in Newtonian dynamics, we can begin to solve some of the world's serious problems. Suppose that, on a windless day, a wellington boot was dropped from a hot-air balloon at a height of 100 m (3281 ft) above the ground. At about what speed would the boot hit the ground? In order to use the equation of motion, we need to know the forces acting on the boot in flight and the mass of the boot. Wellington boots do not usually disintegrate in use, so we may assume that the boot has a constant mass of about 1 kg. Clearly, the force that is bringing the boot down is gravity, so the force acting on the boot is its weight, that is, the mass of the boot multiplied by the constant g, hence $1 \times g$ Newtons. For this rough calculation, the value 10 ms^{-2} can be assumed for g. Let the distance s that the boot has fallen be measured downwards from the balloon and let the boot have velocity v and acceleration a downwards at that point. By the equation of motion,

$$1 \times g = 1 \times a,$$

that is,

$$a = g.$$

Therefore the acceleration of the boot is constant, and we can use the kinematical equation $v^2 = u^2 + 2as$, where u is the initial velocity, the velocity when $s = 0$. But the boot was dropped from the balloon, not thrown, so $u = 0$ and we have the equation

$$v^2 = 2gs.$$

Therefore the velocity V of the boot when it hits the ground is given by $s = 1000$ and $g = 10$, that is, $V^2 = 2000$ and thence $V = \sqrt{2000}$, which is about 141 ms^{-1} or about 316 miles per hour.

At that speed the boot would disappear into the ground without leaving a trace. In fact, if the boot was dropped experimentally, the speed of landing would be found to be much slower than this. So what have we done wrong? If we try throwing a welly in a field, the flapping of the top of the boot will suggest the missing factor: we have ignored air resistance. Our calculation above is a *theoretical model* of the practical experiment, expressing the physical facts in the mathematical terms of Newtonian dynamics, and the failure of our model means only that we have failed to include all the important physical factors. The obvious move is to construct a second mathematical model for the problem, taking air resistance into account.

Air resistance is determined in Newtonian dynamics by the following axiom.

Axiom 8 The air resistance on a particle of mass m moving at speed v less than the speed of sound is a force of magnitude Rv^c acting in the direction opposite to the velocity, where R is a positive constant and c is a constant.

This axiom is well supported by physical experiments, but the experiments are needed to find R and c. The value $c = 1$ is usually correct for low speeds, such as those achieved by throwing, so $c = 1$ would seem to be correct for the descent of our wellington boot. As we cannot perform an experiment just now in order to determine R, let us guess the value $R = \frac{1}{2}$. As the boot is falling (moving in the positive direction for s), the air resistance is acting upwards, so its directed value is $-\frac{1}{2}v$ where v is the velocity. The equation of motion for the second model is then

$$1 \times g - \tfrac{1}{2}v = 1 \times a,$$

that is,

$$g - \tfrac{1}{2}v = a.$$

But our discussion of kinematics also gave the acceleration a as dv/dt, so the equation of motion becomes the differential equation

$$\frac{dv}{dt} = g - \tfrac{1}{2}v.$$

The meaning of this equation is that the rate of change of the velocity is $g - \tfrac{1}{2}v$. When the boot is dropped, $v = 0$, so the velocity changes at g ms^{-2}, so the velocity increases. However, as the velocity v increases, the expression $g - \tfrac{1}{2}v$ decreases, so the rate of increase of v is decreasing. Finally, if v reached the value $2g$ then the rate of increase would be 0, that is, v would no longer increase. We therefore conclude that v will never exceed the value $2g$, therefore the wellington boot will land with a speed not exceeding 20 ms^{-1}, or about 45 miles per hour. Calculations involving calculus show that the speed never quite reaches 20 ms^{-1}, but, for example, the speed reaches 99 per cent of this value in only 163 metres of fall. So the second theoretical model predicts that the boot will land with a speed of about 20 ms^{-1}, and if an experiment disagrees with this value, we know that our guess at the value of R is probably wrong.

The axioms we have considered so far have all been concerned with just one particle, and they are inadequate for systems in which particles are in contact, as can easily occur when the particle consists of the mass of a large body concentrated at the centre of mass. For example, if an anvil is placed on the floor, what force does the floor apply to the anvil?

Axiom 9: The Third Law of Motion The force exerted by a particle A on a particle B is equal but in the opposite direction to the force exerted by B and A, and both forces act in the line AB.

Let us illustrate the use of the Third Law of Motion by means of an example. An elephant is sitting in a lift which is going upwards. What is the force that the elephant exerts on the floor of the lift? Is the force different if the lift is going down? Can elephants fly? To obtain theoretical answers to these questions, let us assume that the elephant has a mass of M kg. By the First Law of Motion (Axiom 4), the lift will remain at rest unless a non-zero total force acts on it and, by the Second Law of Motion (Axiom 5), the force will cause the lift to accelerate. Therefore, at least initially, the elephant will be accelerating upwards at a ms^{-2}. According to the equation of motion, as the mass of the elephant is constant by Axiom 3, the total force acting on the elephant is Ma Newtons. The forces acting on the elephant are its weight Mg Newtons downwards and the reaction of the floor of the lift, exerting R newtons upwards. Therefore we have the equation of motion of the elephant

$$R - Mg = Ma,$$

therefore

$$R = M(g + a) \text{ Newtons.}$$

By the Third law of Motion, the reaction of the floor of the lift on the elephant is equal but opposite to the action of the elephant on the floor. Therefore the elephant exerts a force of $M(g+a)$ Newtons downwards onto the floor. In other words, the force exerted on the floor of the lift can greatly exceed the weight of the elephant when the lift is going up.

The same arguments apply to the lift when it is going downwards, but now the acceleration will be in the opposite direction, so we write it as $-b$ ms^{-2}, where b is a positive number. Then the equation of motion is

$$R - Mg = -Mb,$$

therefore

$$R = M(g - b) \text{ Newtons.}$$

This time the force R which the elephant exerts on the floor is less than its weight. There is no mechanism to pull the lift down, so its fastest downward acceleration would be due to gravity. In this case, the force exerted by the elephant on the floor or by the floor on elephant is $R = 0$. Put differently, it means that the floor is not supporting the elephant at all, so if the elephant takes a dainty skip off the floor of the lift it will rise up and float freely in the air. That is, any elephant can fly by taking a suitable ride in a lift.

Newtonian dynamics has further axioms based on experiments. These refer to other kinds of force, such as frictional forces or the forces exerted by springs, and also deal with other kinds of incident, such as collisions.

Solution

It is high time what we rescued the girl from the frozen pond. According to our deduction from the Second Law of Motion (Axiom 5), the momentum of the mechanical system which comprises the girl as she sits rather sorely in the middle of the pond holding the piece of branch is zero, and must remain so in the absence of forces. In anger at the perfidious tree, the girl throws the broken branch at its trunk and it slides across the ice at a constant speed of u ms^{-1}. Let us regard the direction of the tree from the girl as the positive direction, then the branch has velocity u along that line. Let the girl have mass M kg, the branch have mass m kg and the girl have velocity v ms^{-1}. Then, as the momentum is 0, we deduce that

$$Mv + mu = 0,$$

therefore

$$v = mu/M \qquad \text{ms}^{-1}.$$

In other words, the girl slides, rather more slowly, off the ice on the side of the pond away from the tree. This story has a clear moral: always be ready to take a momentous decision.

4

Useless Mathematics: Arithmetic and Methods of Proof

Mathematicians
Always defend their propositions.
When they're attacked I've often heard 'em
Argue by *reductio ad absurdum*.

Diogenes O'Rell

Arithmetic is the mathematical subject concerned with numbers, especially whole numbers. Its name is derived from the Greek *arithmos*, meaning number. The written history of arithmetic goes back to the earliest documents, although these record only methods and calculations, which must have followed many centuries of oral use of numbers for counting and measuring. The mathematics needed for tending flocks of sheep in the ancient world is limited to skill at counting, but for large engineering projects, such as building a pyramid, there is a need for calculations involving the four arithmetical operations, and for this purpose a suitable set of symbols for the numbers and set of rules for performing the operations on them are required. The Egyptian numerals consisted of symbols for 1, 10, etc., which were entered in sufficient quantity to make up the required sum. Such numbers are easy to add and subtract but require much counting in their interpretation, and they are difficult to multiply and divide. The Babylonians had an excellent system of numerals, which we shall discuss shortly, but they were lost when clay tablets went out of fashion. Later, the large Roman engineering projects called for improved numerals.

The Roman numerals are, in fact, easy to add and subtract and long multiplication can be done systematically, although the process requires subtraction as well as addition. Short division is little worse in Roman numerals than in Arabic, but long division is not easily reduced to a regular procedure.

The superiority of the Arabic numerals does not lie in the qualities of the numerals themselves but in the *place system*, that is, the convention that the position of a digit in a numerical sequence determines the power of ten, the *base*, by which it is to be multiplied. Some of the merits of the place system are independent of the size of the base but, if the positive integer k is chosen for the base, there must be k distinct symbols for the digits (including nought). As adding nought to it does not change a number, there are $k-1$ addition tables of $k-1$ entries to be used in arithmetic with base k and, as multiplying by nought always gives nought and multiplying it by one does not change a number, there are $k-2$ multiplication tables with $k-2$ entries to be used. The Babylonians used a place system with base 60 for their numerals (though they may have written words between the digits) and even extended the notation to the fractional part (corresponding to the decimal part for base ten). In this notation, division by any divisor of sixty leads to a terminating fractional part, but 60 symbols for digits, 59 addition tables and 58 multiplication tables need to be used, and these are unlikely to be easy to read or remember.

Let us compare the Babylonian numerals with the *binary* (or *dyadic*) notation, in which two is the base. There the only digits are 1 and 0, which can be represented by on and off (of a switch), up and down (of magnetism on a tape), or yes and no (to the question 'Is the current flowing?'), and thereby be cunningly incorporated into the design of electronic digital computers. The addition table consists of the one entry $1+1=10$ and there is no multiplication table;

consequently addition is largely concerned with carrying the digit 1, subtraction is largely a process of borrowing and all multiplication and division is long.

The inconvenience of these processes is not the reason why the binary notation is rarely used outside the mechanism of computers. To illustrate this, let us transform the number 25 from decimal to binary notation. This may be regarded as a demonstration of the tedium of changing the base, not as an indication of the merits of the bases used. Let us apply the Division Algorithm to find the remainders of 25 and the successive quotients on division by 2. Then $25 = 2 \times 12 + 1$, $12 = 2 \times 6 + 0$, $6 = 2 \times 3 + 0$ and $3 = 2 \times 1 + 1$. Therefore 25 has the binary expression 11001, meaning $2^4 + 2^3 + 1$. This reveals to us the first disadvantage of the binary notation: 11001 is difficult to read as 25 because our names for numbers are based on powers of ten and these symbols are based on powers of two. Easy reading of binary notation demands names for powers of two which do not have a structure based on the use of ten. As eight and sixteen are satisfactory names (although 'thirty-two' is not and might be called 'fifth' instead), we can read 11001 as 'sixteen, eight and one'. However, 23 new names are needed in order to carry the numerals to the decimal number 10^9 which is required for discussions of national finance, a number of 27 figures in the binary notation. A change to the binary notation would introduce ten-figure numbers to even the sports pages of newspapers.

The everyday use of arithmetical calculations induced teachers to include arithmetic in their curricula since the Roman Empire, and thereby inspired this esctatic poem, which is in the 1570 edition of *The Complete Works of Anonymous*.

> Multiplication is vexation,
> Division is as bad,
> The Rule of Three doth puzzle me
> And Practice drives me mad.

Of course, the *Rule of Three* decrees the following procedure.

To obtain the fourth number which bears the same ratio to the third that the second bears to the first, multiply the second by the third and divide by the first.

But why were arithmetic lessons from the middle ages to the eighteenth century decorated by these delightful rigmaroles? The pupils at such lessons wrote on slates and possessed neither paper nor a textbook – in the earlier period the teacher would also have been without a textbook – therefore all relied on their memories of these prose poems for arithmetical procedures. The maddening 'Practice' in the poem is not that Practice that maketh Perfect, nor yet that Practice that maketh a Lawyer Rich, but the wholly Evil Practice of multiplying Quantities displayed in divers Units, as the succeeding Puzzle doth exemplify.

Problem

A yeoman improves the tilth of a rectangular pyghtle by dressing it with spent fuggles at a rate of 14 520 trugs per acre. The length of the pyghtle is 1 furlong, 2 chains, 2 perches, 3 yards, 2 feet and 3 inches and the breadth is 8 chains, 1 perch, 4 yards, 1 foot and 6 inches. A dray holds 1920 trugsful of spent fuggles and a dray load costs 1 shilling and 7 pence. What is the cost to the nearest farthing of dressing the pyghtle?

(To perform this calculation it is necessary to know that an acre is 10 square chains or 4840 square yards, a furlong is 10 chains, a chain is 4 perches, a perch is 5½ years, a yard is 3 feet, a foot is 12 inches, a pound (£) is 20 shillings, a shilling (s) is 12 pence, a penny (d) is four farthings.)

In the early nineteenth century it would have been expected that the solution of a problem like this would be written out on several pages, starting with a decorated title page and a full statement of the problem, so an arithmetic class of those days would do less than a tenth of the mathematical work in a given time that a modern class would do. The productivity has improved due to better textbooks designed to treat the problems that the writers found in their everyday life. From these sources we learn that a typical textbook writer filled the bath by turning on both taps and removing the plug, and he spent his leisure hours in factories so that he could observe the kind of competition that has been analysed by Stephen Leacock in 'A, B and C – The Human Element in Mathematics'. Some mathematicians are good at practical arithmetic, but the majority regard it as a tiresome bore and consequently are bad at it. This partly explains why, over the centuries, many devices have been invented to facilitate arithmetic: the abacus, chequers (the chequered table cloths formerly used by the Exchequer), logarithm tables, calculating machines and computers.

In order to appreciate the motivation of the developments in arithmetic which transcend the narrow bounds of practical requirements, we need to return in imagination to the lives of the newly numerate shepherds when ancient Egyptian civilization was a futuristic dream. The shepherds are about to cook a stew, so they fill the cooking pot with a quantity of mutton, a number of herbs and some cubed roots, then they balance the pot on a couple of thick posts on each side of the fire. But the pot soon wobbles off the frame of two legs, and the shepherds hastily add two more legs on a line approximately at right angles to the line of the original two. This time the pot is fairly stable and the shepherds say, sheepishly, 'Four legs good, two legs bad.' But why?

After a little time, the local sage mints the concept of *luck*. Clearly, four is *luckier* than two. We can know the lucky from the unlucky numbers by the things with which they are associated. Seven is certainly a lucky number because, in 'Green grow the rushes, O!', 'Seven [is] for the seven stars in the sky', the constellation of the Plough (or Great Bear or Big Dipper) which guides the eyes to the one fixed point in the northern sky which is occupied by the Pole Star, as sought by the sea-fevered mariner who says 'All I ask is a tall ship, and a star to steer her by.' Furthermore, this observation makes it clear that seven is not only lucky but is also the correct number of members for a collection of heavenly bodies. Therefore, after Sir William Herschel discovered Uranus in 1781, the philosopher Georg Hegel argued in a paper published in 1800 that, as the number of the known plants was seven, the harmony of the spheres was complete, and astronomers should cease forthwith from their search for further planets. Unfortunately, Hegel's thesis was too late to reach Giuseppe Piazzi, who discovered Ceres (the first of the 200 or so minor planets) in the same year, and it also encountered a powerful antithesis when the planet Neptune was discovered.

Associational number superstition like this still lives, for example, in the feeling that thirteen (one over the significant dozen) is unlucky, and in the belief of English cricketers that the score 111 (and its multiples) is unlucky for the batsman, presumably because the number resembles the batsman's wicket without the bails. Australian cricketers have no fear of this number, even when written upside down, but regard 87 as unlucky.

Pythagoras collected examples of number superstition, but he realized that the same number can have both good and bad associations. The luck in a number resides in the number's intrinsic properties. Therefore the real reason tht four is lucky is that it is the only number which is the sum and product of the same pair of positive integers, and not because of the potty reason given earlier. Six is also lucky because it is *perfect*, being the sum of its parts (divisors other than the number itself), that is, $6 = 1 + 2 + 3$. On the other hand, Pythagoras and his friends noted that the odd prime 17 came between the square number 16 (that is, 4^2) and the rectangular number 18 (that is, 2×3^2), so 17 is unlucky. Indeed, because the square of side 4 and the rectangle of sides 3 and 6 are the only rectangles with sides of integer length for which the area is numerically equal to the perimeter, the number 17 is exceptionally unlucky. Perhaps this is why the Swedish word for 17 (*sjutton*) is a cuss word. However, the real significance of this result is that it requires a careful proof, not merely a calculation.

On the other hand, 17 seems to be a lucky number

because it is one of the few known primes of the form $2^{2^n} + 1$, where n is a positive integer, which means that, according to a known theorem, a regular 17-agon can be constructed by ruler and compass. Perhaps we should conclude that 16, 17 and 18 are lucky ages to be and the infamy heaped on 17 really belongs to 19, which separates the significant numbers 18 and 20? Or has the concept of luck been defined ambiguously? Is a goalkeeper delighted that lucky 6 goals have been scored against him? Does a batsman at cricket complain if her score ends at the unlucky 222?

Euclid's opinions on luck are not obvious from the *Elements*, but he expounded Pythagorean arithmetical results for their mathematical interest. For example, his discussion of the existence of perfect numbers other than 6 was motivated only by mathematical curiosity.

This brings us to the most extensive part of arithmetic, now called *higher arithmetic* or *number theory* to distinguish to from the arithmetic of buisness and home life. Number theory studies the properties of individual numbers, especially integers, but does not study properties that are associated with the approximate size of the number (which belong to analysis) or the definitions of number system (which belong to set theory and algebra). There is no motivation for number theory apart from curiosity, and no practical applications are expected for the results. It was this aspect that attracted Godfrey Hardy (1877–1947), who feared that any potentially useful mathematics might have an application in welfare, which he abhorred. Indeed, very few results in number theory are of any practical ue, but the subject contains an unusually large proportion of aesthetically pleasing proofs and the methods of proof have been of great importance in other, more practical, parts of mathematics.

For example, an expression which is associated with the positive integer n in much number theory, probability theory and combinatorial theory is $n!$, called *factorial n*, which is given by $n! = 1 \times 2 \times 3 \times \ldots \times (n-1) \times n$. The expression $n!$ is only defined if n is a positive integer and attempts to extend its use for number theoretical purposes led to the following question. Can a continuous function $k(x)$ of the real variable x be defined such that $k(x) = x!$ whenever x is a positive integer? (We shall discuss continuous functions in a later chapter.) Euler constructed the function $k(x)$ in 1729 but, for technical reasons, preferred to use the *gamma function* $\Gamma(x)$ such that $\Gamma(x+1) = k(x)$. The gamma function has many applications in analysis, including its use for the definition of general Bessel function. Bessel functions have many applications in science and engineering, and it has been said that an engineer working without Bessel functions is like a lion-tamer working without a whip. (Of course, that does not imply 'eaten', just that the work would be more harrowing.) The moral is that 'It ain't what you do, it's the way that you do it.'

The name 'number theory' is misleading because there is no central theory in the subject, but only a wide variety of problems and proofs. The problems can be classified according to mathematical themes. For example, one class of problems consists of those concerning just addition. The most famous of these problems is Goldbach's Conjecture, which conjectures that every even number greater than four is the sum of two odd prime numbers. (The primes are not necessarily distinct and the pair of primes need not be unique.) The solution of this problem seems no nearer after 250 years of work. This problem illustrates a feature of number theory: a profound problem can have an elementary statement. Another feature is that two closely related problems may have totally unrelated solutions due to a special feature of a number occuring in one of them allowing a different kind of argument. The following problem concerning addition of integers involves the partitions of an integer k, that is, the different ways in which k can be expressed as a sum of positive integers. For example, the only partitions of 3 are 3, $2+1$ and $1+1+1$.

Problem

Urbania intends to reorganize its currency so that the traditional unit, the *spire*, will in future be divided into 10 *pinnacles*. For reasons of economy, the number N of denominations of coins issued should be the least possible, subject to the condition that every transaction of less than 1 spire should be possible with the use of no more than N coins. What is N? Is there a best possible set of coins for Urbania?

Another addition problem asks 'What are the triangular numbers?'. Problems and solutions in number theory are often based on the discernment of a pattern in the numbers concerned. It is clear that *square numbers* are of the form n^2, where n is an integer, and these can be illustrated by n^2 points arranged with equal spacing in a square with n points on each side. A triangle is obtained by removing all the points above one of the principal diagonals. It will then be noted that the diagonals parallel to this principal diagonal contain successively 1, 2, 3, . . ., n points. Consequently the the nth *triangular number* ι_n is the sum $1 + 2 + 3 + . . . + n$. Therefore $t_1 = 1$, $t_2 = 3$, $t_3 = 6$ (so 6 is triangular as well as perfect – is there another such number?), $t_4 = 10$, $t_5 = 15$ and $t_6 = 21$. Is there a simple formula for t_n in terms of n? In the cases we have listed, the number n divides t_n when n is odd but not when n is even, so perhaps $t_n = \frac{1}{2}nr_n$, where r_n is some suitable expression in n. If this is true then $r_1 = 2$, $r_2 = 3$, $r_3 = 4$, $r_4 = 5$, $r_5 = 6$ and $r_6 = 7$. In other words, the formula $t_n = \frac{1}{2}n(n + 1)$ is correct for $n = 1$, 2, 3, 4, 5, 6. if we calculated t_n and $\frac{1}{2}n(n + 1)$ for some more values of n and found that they were equal, we would only have proved the formula in a finite number of cases, and there would still be an infinite number of cases not accounted for. However, there is a method of proving results about the positive integers, such as our proposed formula $t_n = \frac{1}{2}n(n + 1)$. This method is based on the way we define the positive integers, by starting with 1 and then, at each stage, adding 1 to the largest integer we have formed already in order to form the next one. Consequently, if statement is true for 1 and whenever it is true for a positive integer it is also true for the one above it, then it is true for all the positive integers. This is an informal sketch of the proof of the following principle.

The Principle of Induction   Let $P(n)$ be a statement about the positive integer n.

   If   (a) $P(1)$ is true

   and (b) that $P(n)$ is true implies that $P(n+1)$ is true,

   then $P(n)$ is true for all positive integers n.

It is difficult to give a formal proof of the Principle of Induction because it is not clear where to begin. However, a good starting point for the study of the positive integers has been provided by Giuseppe Peano (1858–1932), but his axioms include the Principle of Induction in slight disguise as Axiom 5. The undefined terms in Peano's Axioms are 'positive integer' and 'successor of a positive integer'. The successor of the positive integer n represents the positive integer $n+1$, but the existence of addition is not being assumed, so the successor is denoted by n' at this stage. In the axioms, the statement $m=n$ means that m and n are the same element in the set of positive integers.

Peano's Axioms for the Positive Integers

Axiom 1   1 is a positive integer.

Axiom 2   For every positive integer n there exists a unique positive integer n' which is the successor of n.

Axiom 3   There is no positive integer whose successor is 1.

Axiom 4   If m and n are positive integers such that $m'=n'$ then $m=n$.

Axiom 5   Let M be a set of positive integers with the following properties: (a) 1 belongs to M

   and   (b) n belongs to M implies that n' belongs to M.

   Then M is the set of all positive integers.

From these axioms, we can start by defining $2 = 1'$, $3 = 2'$ and so forth. That is, we define the names and symbols for the positive integers *recursively* by defining the first and then, at each stage, defining the next one by a formula based on terms that are already defined. Next, we define the sum $m + n$, for the positive integers m and n, recursively by defining $m + 1 = m'$ if $n = 1$ and otherwise, when n is a successor k', we define $m + n = m + k' = (m + k)'$. The usual properties of addition can be deduced from this definition by using the Principle of Induction. Multiplication of positive integers is then defined recursively by the formulae $m \times 1 = m$ and $m \times k' = (m \times k) + m$ (which is just a formalization of the way we usually obtaina multiplication table), and then the basic properties of multiplication can be proved in their turn. Notice that the Peano Axioms can be used to prove all the basic properties of the positive integers, but further axioms are needed if the real numbers are also to be defined. Consequently, those results about integers that can only be proved by using the real numbers cannot be deduced from the Peano Axioms.

As an example of the use of the Principle of Induction, let us now give a proof of the formula that we guessed for the *n*th triangulr number t_n. For this purpose, let $P(n)$ be the statement

$$t_n = \tfrac{1}{2} n (n + 1).$$

Because $t_1 = 1$ and $\tfrac{1}{2} \times 1 \times 2 = 1$, the statement $P(1)$ is true, that is, (a) holds. Let us assume that $P(n)$ is true. (This assumption is called the *induction hypothesis*.) Then $t_n = \tfrac{1}{2} n(n + 1)$. We must now prove $P(n + 1)$, so we consider

$$t_{n+1} = 1 + 2 + 3 + \ldots + n + (n + 1) = t_n + (n + 1),$$

therefore, by the induction hypothesis,

$$
\begin{aligned}
t_{n+1} &= \tfrac{1}{2} n(n + 1) + (n + 1) \\
&= \tfrac{1}{2} n(n + 1) + \tfrac{1}{2}(n + 1)2 \\
&= \tfrac{1}{2}(n + 1)\,(n + 2),
\end{aligned}
$$

by collecting the common factor $\tfrac{1}{2}(n + 1)$. Therefore

$$t_{n+1} = \tfrac{1}{2}(n+1)((n+1)+1),$$

that is, $P(n+1)$ is true. Therefore that $P(n)$ is true implies that $P(n+1)$ is true, so (b) holds. We conclude that, by the Principle of Induction, $P(n)$ is true for all positive integers. Therefore we have proved that, for all positive integers n, the nth triangular number $t_n = \tfrac{1}{2}n(n+1)$.

Let us now consider the part of number theory which is concerned with multiplication. Because division of an integer does not necessarily yield an integer, the most elementary problem about multiplication concerns finding an easy arithmetical test for an integer to be divisible by a given integer, like the tests for divisibility by 2, 3, 5, 9 and 10. One such test is that an integer n is divisible by 11 if and only if 11 divides the difference between the sum of the digits for the odd powers of 10 and the sum of the digits for the even powers of 10 in the decimal expression for 10. For example, for the number 243859 the sums are $2+3+5=10$ and $4+8+9=21$, so 11 divides 243859.

These short methods are useful wherever they apply, but all other cases can be settled with the Division Algorithm. But many problems, such as those involving the possible divisibility of a set of integers by a fixed integer, are more easily solved by using the expressions of the integers as products of prime numbers. This use led Euclid to summarize what was then known about prime numbers and so initiate a branch of number theory which is still vigorous. The obvious first question about prime numbers is 'Is there an infinite number of prime numbers?'. That there is an infinity of integers does not immediately imply that there is an infinity of primes, because there is an infinity of powers of 2 alone. The fact that *infinite* is a negative word, meaning 'not finite', can guide us to the use of *reductio ad absurdum* or *indirect proof* to prove that the number of primes is infinite. In a proof by this method, the statement S to be proved is assumed to be false, then arguments are given to prove something undoubtedly false, such as $0=1$ or, say, Pythagoras' Theorem is false or even that the statement S is true. This contradiction shows that the statement 'S is not true' must be false, therefore S is true.

On the subject of prime numbers, Euclid published the following theorem in the *Elements*.

Theorem

There is an infinite number of positive prime numbers.

Proof

Let us assume that the theorem is false, so there is only a finite number of prime numbers, which we may denote by $p_1, p_2, \ldots, p_n$, where n is a positive integer. Because 2 is a prime number, this list is not empty nd we can define the number $N = p_1 p_2 \ldots p_n + 1$. Then $N > 1$, therefore N is a product of positive prime numbers, which need not be distinct. Therefore there exists at least one prime number q which divides N, therefore N has remainder 0 on division by q. But the only prime numbers are $p_1, p_2, \ldots, p_n$, therefore q is one of these say, $q = p_i$. Then $N = (p_1 p_2 \ldots p_{i-1} p_{i-1} \ldots p_n) q + 1$, therefore, by the Division Algorithm, the unique remainder of N on division by q is 1. Therefore $1 = 0$, which is obviously false. Therefore our hypothesis that the theorem is false has led to a contradiction and we conclude that the theorem is true.

The primes have a very irregular distribution. For example, there are four primes between 190 and 200 but none between 200 and 210. However, there are many results about their distribution, the most notable being the *Prime Number Theorem* which, stated inexactly, says that if x is a large positive number, then the number $\pi(x)$ of primes which do not exceed x is approximately $x/\ln x$, where $\ln x$ denotes the natural logarithm of x. The Prime Number Theorem suggests that the distribution is fairly uniform in broad terms, but the search for *prime pairs*, that is, pairs of primes differing by 2, suggests a very uneven distribution in detail. Known prime pairs range in size from 3 and 5 to $1159142985 \times 2^{2304} \pm 1$, with many examples in between. It is still an open question whether there is an infinite number of prime pairs. If there is an infinite number, then for a given integer n, the smallest gap between two consecutive primes greater than n is 1. But what is the longest gap? We can rephrase that question as 'What is the longest possible sequence of consecutive *composite* (that is, not prime) integers?'. We can answer this question by *construction*, that is, we can construct a set of integers which satisfy the conditions of the following theorem.

Theorem

There are arbitrarily long sequences of consecutive composite integers.

Proof

Let n be any positive integer. Let q be the least prime number such that $q > n+1$. Then $q \geq 3$, so there is at least one prime less than q. Let us define K as the product of all the positive prime numbers less than q. Consider the sequence (C): $K+2$, $K+3$, $\ldots$, $K+q-1$. For each integer j such that $1 < j < q$, there exists a prime number $p \leq j$ such that p divides j. By the definition of K, the prime p also divides K, therefore p divides $K+j$ and $p < K+j$, consequently $K+j$ is composite. This holds for $j = 2, 3, \ldots, q-1$, therefore all the numbers in (C) are composite. But (C) contains $q-2 > n-1$ terms, therefore (C) is a sequence of at least n consecutive composite numbers.

Occasionally mathematics figures in a news item when a new 'largest known prime number' is found. It would be wrong to suppose that this leads to dancing in the corridors of university mathematics buildings, because large primes are sought primarily as a test of the capability of some computer (and as a pleasant test of the computist's ingenuity). Because much is known about numbers of that form, such a prime is likely to be of the form $2^p - 1$, where p is a prime. Euclid started the study of such primes when he proved that if p is a prime number such that $2^p - 1$ is also prime, then $2^{p-1}(2^p - 1)$ is a perfect number. Much later, Euler proved a partial converse: if a positive integer n is an even perfect number then there is a prime number p such that $2^p - 1$ is a prime number and $n = 2^{p-1}(2^p - 1)$. Note that if $2^m - 1$ is a prime number then m must also be a prime number, because if r and s are integers greater than 1, then

$$2^{rs} - 1 = (2^r - 1)\ (2^{r(s-1)} + 2^{r(r-2)}) + \ldots + 2^r + 1),$$

so $2^{rs} - 1$ is not prime. Therefore the construction of the even perfect numbers would be reduced to the discovery of prime numbers if the following conjecture were true:

if p is a prime then $2^p - 1$ is a prime.

We can determine the truth of this conjecture by means of a *counterexample*, that is, an example which satisfies the conditions but not the conclusion of a mathematical satement and therefore shows that the statement is false. In this case, for the prime 11, we have

$$2^{11} - 1 = 2047 = 23 \times 89.$$

We conclude that the conjecture is false. However, $2^{13} - 1$ = 8191 is a prime, so the determination of the even perfect numbers reduces to the problem of finding the primes p for which $2^p - 1$ is prime. On the other hand, no odd perfect number is known, and it has been proved that, if there is one, it is greater than 10^{200}. Further use has been made in our imperfect world more recently because prime numbers are now used in the construction of military codes, an application which would have horrified Godfrey Hardy. As a result, spies are now engaged in smuggling prime numbers, instead of the atoms that used to constitute their contraband.

Number theory, like all mathematics, should be seen as an adventure. The mathematician goes forth on a quest armed with induction, which has infinite range so is stronger than any vorpal sword, with *reductio ad absurdum*, which makes a better cage for the truth than any iron bars, and with mathematical construction, which builds edifices far stronger than stone walls. But the counterexample is an adversary which is stronger than any dragon, either in Komodo or the realms of fantasy.

Solutions

Those interested in agricultural as well as arithmetical practice should realise that, because a yeoman owns his farm, he benefits by improving the soil with a generous quantity of fertilizer, such as 14 520 trugs per acre, that is, 3 trugs per square yard. (Despite its shallowness, a trug basket has considerable capacity.) To find the area of the small field, we calculate its length and breadth in yards, obtaining 278¾ yards as the length and 186 yards as the breadth. The area is therefore 51847½ square yards, and 155542½ trugs of spent fuggles are needed, that is, slightly more than 81 dray loads. It seems unlikely that any brewer would pay his drayman to drive away with his cart only partially filled with used hops, so a possible answer is that the cost is that of 82 dray loads at 1 shilling and 7 pence, that is, 19d per dray. The total cost is therefore 1558d or £6, 9 shillings and 10 pence. However the view may be taken that the cost of fertilizing the pyghtle alone is needed. In that case, the total cost is 1539¼d, to the nearest ¼d, that is, £6, 8 shillings and 3¼ pence. Then, we must suppose, the yeoman puts the remnants of the last dray load of spent fuggles on the midden by the shippon and later uses it for manuring the mickelfield.

The state of the Urbanian economy prohibits further delay in designing their new coinage. As it is impossible to issue one denomination of one coin so that every transaction needs just one coin, let us first assume that the number N of denominations is 2. In order to be able to pay a debt of 1 pinnacle, there must be a 1-pinnacle coin in any system. The partitions of 4 into two parts are $3 + 1$ and $2 + 2$, therefore if $N = 2$ the coins must be for 1 and 2 pinnacles or 1 and 3 pinnacles, but the partitions of 5 into two parts are $4 + 1$ and $3 + 2$, neither of which can be paid with a system of two denominations. Therefore we must try $N = 3$. As there must be a 1-pinnacle coin, payments of 1, 2 and 3 pinnacles can be made with at most 3 coins. To pay 4 pinnacles with at most 3 coins requires the issue of a coin for 2 or 3 or 4 pinnacles. As all amounts from 1 to 9 pinnacles can be paid with at most 3 coins of 1-, 2- and 3-pinnacle denominations, there is at least one possible system of coinage with 3 coins, therefore the required value of N is 3. If we assume that payments of all amounts from 1 to 9 pinnacles are equally likely, then we can measure the efficiency of a coinage system by the total number of ways in which payments of all these amounts can be made. The best system by this criterion has coins of 1-, 2- and 3-pinnacle denominations, with 19 useful combinations. The system with 1-, 2- and 4-pinnacle coins has 17 useful combinations and the system with 1-, 2- and 5-pinnacle coins has 15 useful combinations. There are 10 other possible systems. Note that some other factors affect the choice of small denominations if the large unit of currency is divided into 100 small units.

5

Hidden Snags:
The Need for Proof

'The truth is rarely pure and never simple.'
Oscar Wilde, *The Importance of Being Earnest*, Act I

Mathematics books have the annoying characteristic of starting with long proofs of boringly obvious statements before they start working towards the interesting results. True to this tradition, chapter 4 proved the Principle of Induction before starting the discussion of the irregular distribution of the prime numbers. The readers who require the mathematics for some practical purpose are likely to be particularly infuriated by this mathematical habit because they feel no need for the Principle of Induction, on the grounds that to check the first few cases of a formula seems a good enough proof. Let us explore this point of view by means of an example.

Let n be a variable taking positive integer values and let M be a positive integer. We define

$$a_n = n^5 \sin^2 \frac{(1008!)\pi}{n} + n$$

and

$$S(M) = \sum_{n=1}^{M} a_n = a_1 + a_2 + \ldots + a_n + \ldots + a_M.$$

Is it true that $S(M) = \frac{1}{2}M(M+1)$? The formula holds for all values of M from 1 to 1008, by the formula for triangular numbers, because, for every value of n in this range, n divides 1008!, consequently $\sin (1008!)\pi/n = \sin k\pi$, where k is an integer, so $\sin k\pi = 0$ and $a_n = n$. However, for $n = 1009$, which is a prime number, $(1008!)/1009$ is not an integer, therefore $r = \sin^2 (1008!)\pi/1009 \neq 0$ and $a_{1009} = 1009^5r + 1009$. Therefore

$$S(1009) = 1009^5r + 1009 \times 505.$$

The moral of this example is not that checking 1000 cases is insufficient to prove a formula, but that checking any finite number K of cases is insufficient, as we shall now show. In chapter 4, we proved that there is an infinite number of positive prime numbers, therefore there must be a prime number P which is greater than K. Let us substitute $(P-1)!$ for 1008! in the formula for a_n. This provides a new $S(M)$ which is equal to $\frac{1}{2}M(M+1)$ for $M = 1, 2, \ldots, (P-1)$, where $K \leq P-1$, but not equal to $\frac{1}{2}P(P+1)$ when $M = P$. The Principle of Induction is introduced at an early stage in a book or a course in order to avoid this kind of difficulty.

'But surely', says eager Edgar, 'we can use the first few terms of each side of a proposed equation to determine all the later ones, and if the first few pairs are equal then so are all the later pairs, and the equation must hold.' 'Not so', replies Prudence, 'because we have not proved that the first few terms determine a sequence.' 'Of course it does', says I.Q. Tester. 'Everybody knows that 3, 5, 7 must be followed by 9.' 'Nonsense', says Arithmetica, '3, 5 and 7 are primes, so the next term must be 11. 'Actually, the next term is π', says smart Alec, 'because I say so!' Prudence demurs, 'It can't be anything you like, because it is supposed to fit a formula.' 'The question does not mention one', replies Alec, 'but there is always a formula, in any case. Just look at this theorem.'

Theorem Let $a_1, a_2, a_3, \ldots, a_k$ be a finite sequence of numbers. Choose an arbitrary sequence (perhaps infinite) of numbers as a_{k+1}, a_{k+2}, a_{k+3}, ... , a_n, ... Then there is a function b_n of the positive integer variable n alone, defined by a formula, such that $b_n = a_n$ for all the values of n for which a_n is defined.

Proof For a positive integer n and a positive integer $k \leq n$,

$$\delta_n(k) = \frac{(k-1)\ (k-2)\ \ldots\ (k-n-1)}{(n-1)!}.$$

Then $\delta_n(k) = 0$ for $k = 1, 2, 3, \ldots, n-1$ and $\delta_n(n) = 1$. Now let

$$b_n = \sum_{k=1}^{n} \delta_n(k)\ a_k,$$

for all the values of n for which a_n is defined. Then, because of the value of $\delta_n(k)$, for all thee values of n, we have $b_n = a_n$.

In chapter 2 we found the positive integer solutions of $x^2 + 5y^2 = K$ in the case $K = 41$ by using the integral domain D of all numbers of the form $a + b\sqrt{-5}$, where a and b are integers. Let us now apply the same technique to the case $K = 9$. The number 9 factorizes as 3×3, and hence also as -3×-3, in D and it can be shown that 3 is irreducible in D; so, as before, our equation can be rewritten as

$$x + y\sqrt{-5} = A$$
$$x - y\sqrt{-5} = B.$$

From the factorization of 9 into irreducible factors, the numbers A and B are both real, and therefore $y = 0$, so there is no set of positive integer solutions of $x^2 + 5y^2 = 9$. So we have again triumphed by means of the integral domain D. Still, it is better that we should note that

$$(2 + \sqrt{-5})\ (2 - \sqrt{-5}) = 2^2 - (\sqrt{-5})^2 = 9,$$

and therefore the equation $x^2 + 5y^2 = 9$ has the solution $x = 2$ and $y = 1$, an improper possession for an equation with no set of positive integer solutions. Where have we gone wrong this time? Our assertion that the numbers A and B in the equations must be real was based on the assumption that the only products formed by multiplying the units 1 and -1 and the irreducible divisors of 9 were real, and this relied on the assumption that a set of irreducible factors of 9 in D is unique except for unit factors. Is it really unique? It can be shown that $2 + \sqrt{-5}$ and $2 - \sqrt{-5}$ are both irreducible in D, therefore the equations

$$3 \times 3 = 9 = (2 + \sqrt{-5})\ (2 - \sqrt{-5})$$

prove that the factorization of a number in D into irreducible factors in D is not necessarily unique except for unit multipliers.

Is the factorization of an ordinary integer as a product of irreducible integers essentially unique? Until Gauss discovered the non-uniqueness of factorization in the integral domain D discussed above, nobody had realized that the uniqueness required a proof, but Euclid had supplied the basis for the proof in the *Elements* when he gave the process for calculating a greatest common divisor now called the *Euclidean Algorithm*. Let us use this process to find the greatest common divisor of 9198 and 10439. By the Division Algorithm,

$$
\begin{aligned}
10439 &= 9198 + 1241, \\
9198 &= 7 \times 1241 + 511, \\
1241 &= 2 \times 511 + 219, \\
511 &= 2 \times 219 + 73, \\
219 &= 3 \times 73.
\end{aligned}
$$

Because an equation $a = zb + c$ implies that the greatest common divisor of a and b is also the greatest common divisor of b and c, the fact that 73 divides 219 implies that 73 is the greatest common divisor of 10439 and 9198. Furthermore, these equations can be used to express the number 73 in terms of 10439 and 9198 as follows.

$$
\begin{aligned}
73 &= 511 - 2 \times 219 \\
&= 511 - 2 \times (1241 - 2 \times 511) \\
&= 5 \times 511 - 2 \times 1241 \\
&= 5 \times (9198 - 7 \times 1241) - 2 \times 1241
\end{aligned}
$$

$$= 5 \times 9198 - 37 \times 1241$$
$$= 5 \times 9198 - 37 \times (10439 - 9198)$$
$$= 42 \times 9198 - 37 \times 10439.$$

Indeed, this process can be used to prove that if a and b are non-zero integers with greatest common divisor d, then there exist integers s and t such that $sa + tb = d$. Euclid used this result to prove the following.

Theorem

Every irreducible integer is prime.

Proof

Let a be an irreducible integer and let a divide the product bc of the non-zero integers b and c. If d is a greatest common divisor of a and b then d divides a, so there exists an integer x such that $a = xd$. Because a is irreducible, either x is a unit or d is a unit. If x is a unit then x divides every non-zero integer and, as d is the greatest common divisor of a and b, d divides b, therefore x divides b/d, and hence $a = xd$ divides b. Otherwise, d is a unit in the integers and so is 1 or -1. By changing the sign of c, if necessary, we may assume that $d = 1$. Then, by the Euclidean Algorithm, there exist integers s and t such that $sa + tb = 1$, therefore $(sc)a + t(bc) = c$. Because a divides both bc and itself, a divides c. We conclude that, whenever the irreducible integer a divides bc it divides at least one of b and c, that is, a is *prime*.

It is easy to prove that, in any integral domain, every prime element is irreducible and therefore the terms 'prime' and 'irreducible' can be used synonymously for integers. Gauss used induction and the theorem above to prove that the factorization of an integer into a product of a unit and positive irreducible integers is unique. However, the comment 'That's obvious' seems far less appropriate that the remark of Gauss's contemporary, the Duke of Wellington, who said 'It was a damned close run thing.' Perhaps now it will be easier to understand why mathematicians start the account of any topic by proving basic results which are as

71

obvious as $2+2=4$. But then is $2+2=4$? The only difficulties in proving this result in its usual sense concern the definition of the set of integers **Z** and the proof that **Z** is an integral domain. If we change notation to, let us say, base four, then we have $2+2=10$, but 10 is merely four in disguise. The problem is whether there is another sensible arithmetic in which $2+2$ is not four. Perhaps we ought to ask first whether there is any arithmetic other than that for the rational numbers, real numbers and complex number based on the properties of **Z**.

The passage of time will make it obvious that there are other possible, and entirely practical, arithmetics. Suppose that it is 9 o'clock now, then in 6 hours' time it will be 3 o'clock; that is, in timely arithmetic, $9+6=3$. But what about 33 hours later? We add 33 to 9, then use the Division Algorithm to find the remainder and we discover that in 33 hours it will be 6 o'clock. We can simplify that calculation by applying the Division Algorithm to find the remainder 9 of 33 on division by 12 then, in timely arithmetic, $9+9=6$. The process we use in adding times given in hour is to add the numbers and then take the remainder on division by 12. The sum of two numbers a and b in timely arithmetic is not altered by adding multiples of 12 to them or to their sum because the Division Algortihm ensures that the remainders of a, b and $a+b$ are uniquely defined.

Not only do we have ordinary arithmetic and timely arithmetic, we are also acquainted with daily arithmetic. Suppose that today is Tuesday. What day is 39 days later? Let Sunday be 1, Monday 2, Tuesday be 3, and so on. now add 3 and 39 to obtain 42. Divide by 7 and find the remainder 0. So the day will be a Saturday. Daily arithmetic is the same as timely arithmetic but with 7 substituted for 12. We can also multiply timely or daily numbers, though this has no immediately obvious application. For example, in daily arithmetic, the product of 3 and 6 is 4 because $18=2\times7+4$.

Gauss introduced arithmetic of *congruence classes modulo m* for every integer m greater than 1 with addition and multiplication that gives consistent results in all cases. A special notation is not required for these numbers, but simply a distinctive notation for the 'equality' relation. We say that the integers a and b are *congruent modulo m* if and only if m divides $a - b$, and we write this as $a \equiv b \pmod{m}$. This statement is equivalent to saying that a and b have the same remainder on division by m. Notice that, although the use of any of the integers is permissible with this notation, there is no need to use any except $0, 1, 2, \ldots, m-1$. For example, if $m = 2$, the only remainders are 0 (for even numbers) and 1 (for odd numbers) with relations such as $1 + 1 \equiv 0 \pmod{2}$. The notation for integers modulo m can also be helpful with many traditional types of arithmetic problem, such as the following.

Problem

In order to improve the royal finances, the King of Ruritania decreed that if anybody over the age of 60 died while owning more than four manors (villages), including the estates of all unmarried daughters, then the ownership of all his or her manors would pass to the king. By the operation of this law, the king's financial position improved and the number of manors beyond the king's ownership was reduced to 83. Baron von Haerdupp decided to maintain his family's wealth by distributing his manors equally among his five sons on his 59th birthday, keeping the remaining four for himself. However, shortly before his birthday, his eldest son was killed in a duel when he was defending the honour of the Crown Prince, so the baron had a deed of transfer drawn to divide his manors among his four sons, keeping the remaining three for himself. At this moment, the king honoured the family by offering the princess's hand in marriage to the baron's second son. To avoid some of his manors falling into the hands of the royal family, the baron then divided his manors among his three other sons and kept for himself just Schloss Haerdupp and the largest of the country manors. How many manors did the baron own initially?

Arithmetic modulo an integer can facilitate many ordinary arithmetical operations, such as investigating whether

$$2^{82}(2^{83} - 1)$$

is perfect. A theorem of Fermat implies that a proper prime divisor q of $2^{83} - 1$ must have the form $q = 166n + 1$, where n is a positive integer and $q < 2^{41}$. Therefore we can start by finding the remainder of $2^{83} - 1$ on division by 167. For this purpose, we calculate modulo 167 as follows.

$$2^8 \equiv 256 \equiv 89 \pmod{167}$$

therefore $2^{16} \equiv 89^2 \equiv 7921 \equiv 72,$

therefore $2^{32} \equiv 72^2 \equiv 5184 \equiv 7,$

therefore $2^{64} \equiv 7^2 \equiv 49,$

therefore $2^{67} \equiv 49 \times 8 \equiv 392 \equiv 58,$

therefore $2^{83} \equiv 2^{67} \times 2^{16} \equiv 58 \times 72 \equiv 4176 \equiv 1,$

therefore $2^{83} - 1 \equiv 0 \pmod{167}.$

We conclude that 167 divides $2^{83} - 1$, which is therefore not a prime.

We can now answer the question as to whether $2 + 2 = 4$. In fact,

$$2 + 2 \equiv 1 \pmod 3 \text{ and } 2 + 2 \equiv 0 \pmod 4$$

are entirely practical alternatives for $2 + 2$. If the answer 5 is insisted upon, we can have

$$2 + 2 \equiv 5 \pmod 1,$$

but that equation really means $0 + 0 \equiv 0 \pmod 1$, which is, perhaps, not very exciting.

Solution

Finally, let us enumerate Baron von Haerdupp's manors. Although we are using Gauss's ideas in our calculations, this kind of problem was solved independently by Roman and Chinese mathematicians in about 100 AD, and consequently the following result is called the *Chinese Remainder Theorem*.

Let n be a positive integer, let $a_1, a_2, \ldots, a_n$ be integers and let m_1, m_2, . . . , m_n be integers greater than 1 such that the greatest common divisor of any pair of them is 1. Then the equations $x \equiv a_i$ *(mod m_i), for* $i = 1$, 2, . . . , n, have a common solution which is unique modulo $m_1 m_2$ . . . m_n.

We shall not need the Chinese Remainder Theorem in our solution. The information concerning the number of manors which remain after sharing out the others equally gives us the following congruences which are satisfied by the number x of manors that Baron von Haerdupp owns initially.

$$x \equiv 4 \ (\text{mod } 5),$$
$$x \equiv 3 \ (\text{mod } 4),$$
$$x \equiv 2 \ (\text{mod } 3).$$

From the first congruence, there exists an integer y such that $x = 5y + 4$, so the second congruence gives us

$$5y + 4 \equiv 3 \ (\text{mod } 4),$$

from which follows

$$y \equiv 3 \ (\text{mod } 4).$$

Therefore there exists an integer z such that $y = 4z + 3$ and hence $x = 20z + 19$. Then, from the third congruence,

$$4z \equiv 2 \ (\text{mod } 3)$$
therefore $\quad\quad z \equiv 2 \ (\text{mod } 3).$

We conclude that there exists an integer t such that $z = 3t + 2$, and therefore $x = 60t + 59$. But x is negative if t is negative, so t is non-negative. If $t = 0$ then $x = 59$ but if t is positive then $x \geq 119 > 83$, the total number of manors not owned by the king. Therefore Baron von Haerdupp started with 59 manors. Sad to relate, his story became tragic after he gave 57 of them to his sons, because he had to mortgage his other two manors heavily in order to be able to provide dowries for his two stepdaughters, and this brought such poverty to his family that his pretty daughter was virtually reduced to the status of a domestic servant.

6

The Space between the Numbers: Geometry

'Not five yards from the mountain path
This thorn you on your left espy;
And to the left, three yards beyond,
You see a little muddy pond
Of water, never dry;
I've measured it from side to side
'Tis three feet long and two feet wide.'

William Wordsworth, 'The Thorn', first edition

The impact of tragedy is intensified by precision, as Wordsworth demonstrated in this immortal stanza from 'The Thorn', which also reveals that he appreciated that geometry includes the useful art of surveying, as indicated by its derivation from the Greek *ge* (earth) and *metron* (measure). However, unless he learned about it at university or from his friend Sir William Rowan Hamilton, Wordsworth probably knew little of the remarkable work on plane sections of a cone by Apollonius of Perga (260–200 BC) because, until this century, Euclid's voluminous *Elements* left no space in the school syllabus for any other geometry. Apollonius wrote *Conics* in order to give a geometrical construction for the cube root of 2, to answer a problem set by an oracle. he solved the problem by means of the intersection of some sections of a cone – brilliant mathematics even when viewed with hindsight through coordinate geometry. As we can now extract cube roots with our pocket calculators, the subsequent value of *Conics* lies in its classification of conic sections, which has been applied to astronomy and was the basis of the books on ballistics written before the discovery of calculus.

A *cone* in three-dimensional geometry is the set of points which lie on the lines joining a fixed point, the *vertex* V, to a closed curve. As the lines extend infinitely in both directions, a general cone has the appearance of an ill-made, doubly infinite wigwam. Here we shall use the word *cone* in its usual sense, to mean a *right circular cone* K, though our cones extend infinitely in both directions. According to Apollonius, a *conic* S is the curve of intersection of K with a plane P. If P passes through the vertex V, the conic S consists of a pair of lines or one line (which should be counted twice because in this case P is the tangent plane to K at every point of the line) or just the point V. We call these *degenerate conics*. We shall now assume that P does not pass through V, so the conic we obtain is *proper*. If P is perpendicular to the axis of the cone (its line of symmetry), then S is a circle. if P meets the axis at an angle which is greater than the half-angle at the vertex of the cone, then S is a closed curve which is called an *ellipse*. If P is parallel to one of the lines in the surface of K, then s is open at one end and is called a *parabola*. If P meets the axis at an angle which is smaller than the half-angle of the cone, then P intersects both ends of K and S is the curve called a *hyperbola*, which has two open parts.

Unfortunately, the other Greek geometers did not regard Apollonius' extraction of the cube root of 2 as a proper solution to the problem, and they continued to look for a construction which used only a straight edge and compass acting in one plane. They also attempted two other ruler and compass constructions: to trisect a given angle and to 'square the circle', that is, to construct a square of the same area as a given circle. Their attempts have earned them everlasting respect, because they failed completely. And all three constructions are impossible. In 1882, F. Lindemann proved that π is the root of no polynomial with integer coefficients, and this shows that a line segment of length π units cannot be obtained by a geometrical construction. In the nineteenth century, the algebraic methods of Evariste Galois were used to show that the cubic equations $x^3 = 2$ and $x^3 - 3x - 1 = 0$ (which represents the trisection of an angle of $\pi/3$ radians) cannot be solved using only square roots and other roots of order a power of 2, which are the only kinds of root that can be constructed using a ruler and compass.

The Roman mathematicians took more interest in conics and produced the following definition of a proper conic, which was proved to be equivalent to Apollonius' definition by Pappus of Alexandria during the fourth century AD and was shown to be of great astronomical importance by Johan Kepler (1571–1630). Take a fixed point F, the *focus*, a fixed line L, the *directrix*, and a positive real number ε, the eccentricity. Then the locus S of points for which the distance from F is ε times the distance from L is a proper conic. With this definition, S is an ellipse if $\varepsilon < 1$, a parabola if $\varepsilon = 1$ and a hyperbola if $\varepsilon > 1$. The circle corresponds to the limiting case where $\varepsilon = 0$.

Pappus also proved a theorem which has an unusual feature: it is concerned only with points and lines and it statement involves neither length nor angle. We shall discuss the significance of this later in the chapter.

Pappus' Theorem Let L and M be two lines in a plane, let A_1, A_2 and A_3 be points on L, let B_1, B_2 and B_3 be points on M, let P be the intersection of A_2B_3 and A_3B_2, let Q be the intersection of A_3B_1 and A_1B_3, and let R be intersection of A_1B_2 and A_2B_1. Then P, Q and R are collinear, that is, lie line on one line.

Leonardo da Vinci (1452–1519) and Albrecht Dürer (1472–1528) arranged the size of figures and objects in their pictures according to *perspective* in an attempt to make flat images resemble the three-dimensional scenes they depicted, whereas medieval painters adjusted the sizes of the figures and objects to indicate their importance in the picture. With customary speed, mathematics came to the aid of art only about a hundred years later with a first discussion of the geometry of perspective in the work of Girard Desargues (1593–1662) and Blaise Pascal (1623–1662). We can understand their basic idea of *conical projection* if we imagine a one-eyed artist painting a landscape on a plane sheet of glass. When he places his eye in front of the centre of the glass and looks along some curve in the landscape, the rays of light meet it. The image of a straight line in the landscape is always a straight line and the image of a circle on a plane parallel to the glass with its centre in line with the centre of the glass is a circle.

However, as Apollonius' definition of conics makes clear, the images of other circles are conics and the images of some conics are circles. Because the image of a circle can be an ellipse, it follows that equal lengths on different lines need not have equal images, but the lengths of images of segments of one line are related to the original lengths. In order to explain this, we need the idea of the *cross ratio* (A,B;C,D) of the points A, B, C, D on a line, which is given by

$$(A,B;C,D) = \frac{AC \times DB}{CB \times AD,}$$

where AC represent the directed length of the line segment from A to C. If the rays of light from A, B, C, D meet the sheet of glass in the points A′, B′, C′, D′ then it can be proved that

$$(A,B;C,D) = (A',B';C',D').$$

Suppose that the point C lies between A and B such that AC/CB = k. What can we say about A′C′/C′B′? If the line AB meets the plane of the glass in the point D, we may take D′ = D and obtain (A,B;C,D) = (A′,B′;C′D). Therefore

$$\frac{A'C'}{C'B'} = k\frac{DB \times A'D}{AD \times DB'}.$$

This formula led Desargues to the idea of the *point at*

infinity I on the line AB, a notional point on AB such that any line parallel to AB meets AB at I. (Kepler also invented points at infinity at about the same time.) Desargues not only investigated conical projection, but he also proved a theorem relevant to perspective which involves only lines and points, just like Pappus' Theorem. Pascal's main contribution to geometry was the following theorem about conics, of which Pappus' is the baby version concerning a degenerate conic.

Pascal's Mystic Hexagram Let S be a conic and let A_1, A_2, A_3, B_1, B_2, B_3 be points on S. Let P be the intersection of A_2B_3 & A_3B_2, let Q be the intersection of A_3B_1 & A_1B_3 and let R be the intersection of A_1B_2 & A_2B_1. Then P, Q & R are collinear.

The diagram for this theorem appears in figure 1. The mystic aspect of the theorem is not entirely obvious, but pherhaps it resides in the fact that this intellectually symmetrical figure has six vertices, and six is a perfect number. Therefore Pascal had a perfect figure, coveted by so many nowadays. However, it was easy for Pascal because he was only sixteen when he proved the theorem. In addition, he made use of the fact that any image of a mystic hexagram under conical projection is also a mystic hexagram, in an essay on conics in which he derived Apollonius' theorems as special cases of the mystic hexagram theorem.

Desargues and Pascal were unlucky because attention was drawn away from their work almost immediately by the revolution instigated by that most gentlemanly and philo-sophical of anarchists, René Descartes, who introduced the *analytical* method into geometry to augment the traditional *synthetic* method. By choosing a pair of lines at right angles in the Euclidean plane as coordinate axes, Descartes was able to represent a point by the ordered pair of real numbers (*Cartesian coordinates*) which are the distances of the point from the axes. He could then represent a curve as either the graph of a function or as a set of points for which the coordinates satisfy an equation.. This immediately allows a

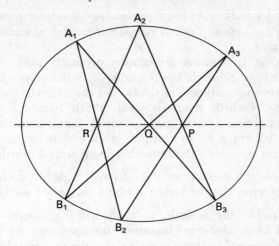

Figure 1 Pascal's Mystic Hexagram

discussion of cubic curves and other curves of higher degree and it is also the first step towards the invention of geometries of dimension greater than 3. Moreover, the analytical method also paved the way for the greater geometrical revolutions of the nineteenth century.

However, in accordance with Newton's Third Law, there was a reaction by traditionalists, who suggested that the analytic method was not valid because it used theorems (in algebra) not derived from Euclid's axioms. Although this objection does not apply to analytical geometry (which merely makes use of the length in synthetic geometry), this can be a valid objection to the application of one branch of mathematics to another. The traditionalists also had the conviction (equally false, as we observed in chapter 2) that the arguments of synthetic geometry were the only rigorous ones. It was in order to avoid this controversey that Newton expressed all his work on calculus in the language of synthetic Euclidean geometry, though he probably obtained his results by analytical methods. Much later, the controversy between synthetic and analytical geometers was aggravated by the rivalry of warring nations to the extent that some

French journals would not publish any synthetic geometry and some German journals would not publish any analytical geometry.

Happily, later nineteenth-century geometry made it clear that the best approach was to employ a judicious mixture of synthetic and analytical methods, but there remains a belief that the synthetic methods yields short, beautiful proofs whereas the analytical method yields long, uninstructive proofs burdened by discussions of exceptional cases. This sentiment overlooks the following fundamental principle.

The Law of Conservation of Difficulty   A difficulty can be transformed but not removed by a change of method.

For example, the difficulty in an analytical geometry proof created by a coefficient becoming nought reappears in the synthetic proof as two items in the figure coinciding or becoming parallel, which calls for a fresh start in the synthetic proof with a new figure.

Euclidean geometry may be called *Cartesian geometry* when it is treated analytically by means of Descartes' system of coordinates. The point with coordinates (x, y) is on the line L in Cartesian geometry if and only if $ax + by + c = 0$, for suitable constants a, b and c determined by L. That is, $ax + by + c = 0$ is an *equation* for L. By using Pythagoras' Theorem, it can be proved that the distance between the point with coordinates (x, y) and the point with coordinates (u, v) is

$$\sqrt{(x - u)^2 + (y - v)^2}.$$

Therefore the point with coordinates (x, y) is on the circle with centre (u, v) and radius r if and only if

$$(x - u)^2 + (y - v)^2 = r^2,$$

so this is called an *equation* of the circle. In Cartesian geometry, a conic can be identified as the set of all points which satisfy an equation of degree 2.

To define Cartesian coordinates for three-dimensional space, we start by choosing coordinate axes 0x and 0y in one plane, the *xy-plane*. Through the *origin* 0, which has coordinates $(0,0)$ in the xy-plane, we choose a positive direction on a line at right angles to form the z-axis $?z$. (Actually, there is a rule which determines the direction to be chosen.) The two axes $0z$ and $0x$ lie in the *zx-plane* and the axes $0y$ and $0z$ lie in the *yz-plane*. For a point P in space we find the directed distances x, y, z of P from the yz-, zx- and xy-planes and the set of coordinates of P (with respect to the chosen axes) is the ordered triad (x,y,z). The distance of P from the point with coordinates (a,b,c) is given by

$$\sqrt{(x-a)^2 + (y-b)^2 + (z-c)^2}$$

and planes are represented by linear equations

$$px + qy + rz + s = 0,$$

where p, q, r, s are real numbers.

A single equation in the coordinates represents the condition that P should lie on a surface, therefore two equations are needed to define a curve as the intersection of two surfaces. For example, a line can be determined by the intersection of two planes, and therefore by two linear equations. In order to obtain an informative and symmetrical pair of equations for a line L, we note that L is defined by the coordinates (a,b,c) of a point a on it and the direction of L, which can be expressed as the coordinates (l,m,n) of a point B such that OB is parallel to L. Then the point P with coordinates (x,y,z) is on L if and only if the direction AP is parallel to OB, which holds if and only if the point with coordinates $(x-a, y-b, z-c)$ is on the line OB, that is, if and only if

$$\frac{x-a}{l} = \frac{y-b}{m} = \frac{z-c}{n}.$$

These equations for the line L are really a way of writing the equations of two planes so that the direction of L and a point on L can be read easily from them. The idea of defining a direction by the coordinates of a point is used in the technique of *vector analysis*, which has very important applications in theoretical mechanics. The direction of L is best indicated by the point with coordinates (u,v,w) which is a unit distance from O on OB. We call $\mathbf{m} = (u,v,w)$ the *unit vector* in the direction of OB and we call $\mathbf{r} = (x,y,z)$ the *position vector* of P. Vectors (a,b,c) and (f,g,h) are added by the rule

$$(a,b,c) + (f,h,h) = (a+f, b+g, s+h),$$

which corresponds to the law for adding forces in mechanics. There are two kinds of multiplication in vector analysis. For a force $\mathbf{D} = (F,G,H)$ and the unit vector $\mathbf{m}$, the *scalar product*

$$\mathbf{m} \cdot \mathbf{D} = uF + vG + wH$$

is just a number, which represents the component of the force $\mathbf{D}$ in the direction of $\mathbf{m}$. The *vector product*

$$\mathbf{r} \times \mathbf{D} = (yH - zG, zF - xH, xG - yF)$$

is a vector which represents the moment about O of a force $\mathbf{D}$ acting at a point P with position vector $\mathbf{r}$. With these operations, vector analysis provides a brief notation for mechanics which expresses the mechanical concepts clearly while reducing the emphasis on the computational details.

Cartesian geometry is Euclidean, irrespective of the number of dimensions, so we still have not proved the existence of a non-Euclidean geometry. A particular form of the problem is this: if a geometry satisfies Euclid's Axioms 1, 2, 3 and 4 as listed in chapter 2, does the geometry also satisfy Playfair's Axiom (or, equivalently, the complicated Axiom 5)? There is a simple counterexample, which should have been discovered by Christopher Columbus (1445–1506), along with America and the fact that the West Indies are not part of China. Columbus failed to find America because he was not looking for it, but, perhaps, we can find a non-Euclidean geometry by studying two aspects of North America.

First, let us consider the journey from Winnipeg to Hecla Provincial Park on Lake Winnipeg by the direct route on Provincial Trunk Road 8. The road goes directly north for 51 km through countryside divided into mile squares by the government surveyors, then it doglegs to the east before going directly north for a further 54 km. Why does the road dodge like that? The local explanation is that the dogleg is at a correction line – apparently the surveyors could not even divide the countryside into squares without making mistakes. Was their geometry faulty, or were the distracted by the wildlife? The latter is possible because the fauna of Manitoba includes the bison, the wolf and two kinds of bear. And that brings us to the second aspect of North American geometry, a problem which applies geometry to hunting.

Problem

A bear hunter saw a large bear feeding to her north. The bear was comfortably in range, but the line of fire was obstructed, so the hunter stalked 500 m to the west until she had a clear view of the bear. The bear had not seen her and was still feeding at the same place. The hunter fired her rifle due north and killed the bear. What colour was the bear?

We shall appreciate the Manitoba surveyors' difficulties rather better if we consider the geometry R of the surface of the earth, which we shall assume to be the surface of a sphere. There are no lines on R, but their property of supplying the shortest route between two points in the plane belongs in r to the *great circles*, that is, the circles which have as their centres the centre of the sphere. The other circles on the surface of a sphere are called *small circles*, and they have properties in R like those of circles in the plane. It is easy to see that, with these definitions, the geometry R satisfies Euclid's Axioms 1, 2, 3 and 4. In order to determine the truth of Playfair's Axiom in R, let us consider two distinct lines L and M in R, that is, the great circles L and M of the sphere. As L and M both lie in planes which contain the centre of the sphere, the planes meet in a line (in Euclidean space) which is a diagonal of the sphere and hence a diagonal of both the circles L and M. Therefore the 'lines' L and M in R meet. (They actually meet in two points, but the definition of R can be modified to give an abstract geometry in which the lines meet only once.) Therefore the geometry R satisfies the following axiom.

Riemann's Axiom Any two distinct lines meet in at least one
point.

We have constructed the non–Euclidean geometry R inside
Euclidean three-dimensional geometry, therefore we have
shown that if Euclidean geometry exists, then so does non-
Euclidean geometry.

Arthur Cayley created algebraic geometry by applying the
methods of algebra to a combination of the geometrical ideas
of Desargues and Descartes. This is an abstract geometry
which can either be constructed starting with numbers or
defined by axioms about points and lines. However, instead
of defining algebraic geometry properly, we shall derive it
informally from Cartesian geometry, and we start by
defining a slightly more general system of coordinates.

Suppose that we have already chosen a system of
Cartesian coordinates in which the general point P has the
coordinates (X, Y). For any non-zero real number z, we
write $x = Xz$ & $y = Yz$ and we form the ordered triad of real
numbers (x, y, z), which is a set of *homogeneous (rectangular)
Cartesian coordinates* for P. Notice that P does not have a
unique set of homogeneous Cartesian coordinates; for
example, one set is $(X, Y, 1)$, another is $(2X, 2Y, 2)$. However,
any two sets of coordinates for P are proportional, and any
set of coordinates proportional to $(X, Y, 1)$ is a set of
coordinates for P. This is not a serious disadvantage, because
in the analogous case concerning lines we readily recognize
$Y = \frac{1}{2}X + 1$, $\frac{1}{2}X - Y - 1 = 0$ and $-X + 2Y + 2 = 0$ as equations of the same line, which has the equation
$-x + 2y + 2z = 0$ in homogeneous Cartesian coordinates. In
fact, when homogeneous Cartesian coordinates are used, the
equation of any (algebraic) curve has every term with the
same degree in x, y & z (taken together). For example, a
circle has an equation of the form

$$x^2 + y^2 + 2fyz + 2gzx + cz^2 = 0.$$

The symmetry of these *homogeneous* polynomials makes
many calculations easier, but lengths and angles are more
awkwardly expressed. However, there are two asymmetrical

features of the algebra of homogeneous Cartesian co-ordiantes: (x,y,z) (where not all x,y,z are 0) represents a point if and only if $z \neq 0$, and $ax + by + cz = 0$ (where not all a,b,c are 0) represents a line if and only if it is not $cz = 0$. Let us remove these anomalies. Let the set of all ordered triads of real numbers proportional to $(x,y,0)$, where x & y are not both 0, be the coordinates of a *point at infinity* and let the equation $z = 0$ represent the *line at infinity*. Notice that a point is at infinity if and only if it is on the line at infinity.

By adding these notional points and line to the Cartesian plane, we have created an abstract geometry which contains the Cartesian geometry. In this larger geometry the lines with equations $ux + vy + wz = 0$ and $lx + my + nz = 0$ (where neither is the line at infinity) meet in the point at infinity $(a,b,0)$ if and only if $ua + vb = 0 = la + mb$, which holds if and only if $u/v = l/m = -b/a$, that is, the two lines have the same gradient. In other words, to finite lines meet at a point at infinity if and only if they are parallel. Consequently, in our large geometry, any two lines meet at a point, but the point is at infinity if the lines are parallel.

Finding the intersections of curves other than lines introduces some difficulties which we shall indicate by the following examples. We start by considering the points of intersection of the cubic curve K with the equation $x^3 - y^3 - 2y^2z + z^3 = 0$ and the circle C with the equation $x^2 + y^2 - 4z^2 = 0$. The equation of K gives us $x^3 = y^3 + 2y^2z - z^3$ and the equation of C gives us $x^2 = 4z^2 - y^2$. We can eliminate x from these equations by equating the two values of x^6 and so obtaining

$$(y^3 + 2y^2z - z^3)^2 = (4z^2 - y^2)^3.$$

If we write $Y = y/z$, expand these brackets and collect terms, we can obtain the equation

$$2Y^6 + 4Y^5 - 8Y^4 - 2Y^3 + 44Y^2 - 63 = 0.$$

If we can solve this equation we can immediately determine the (at most) six points of intersection of K and C but, unhappily, we have no technique for finding an exact solution of this equation, and an approximate solution is unsuitable for geometrical purposes, such as testing whether points are collinear. However, if we only require the points of intersection for a further construction, there is a device for circumventing this difficulty.

Suppose that we have two curves R and S with equations $M=0$ and $N=0$. For any real numbers μ and ν (not both 0), the curve T with equation $\mu M + \nu N = 0$ passes through all the points of intersection of R and S. If R and S are both lines, then the equation $\mu M + \nu N = 0$ is of the first degree, so T is a line through the intersection of R and S. Similarly, if R and S are conics then T is also a conic. In the special case where R and S are the circles with the equations

$$M \equiv x^2 + y^2 + 8xz + 12z^2 = 0$$

and

$$N \equiv x^2 + y^2 - 2xz = 0,$$

the equation of T is

$$(\mu + \nu)(x^2 + y^2) + (8\mu - 2\nu)xz + 12\mu z^2 = 0.$$

The conic T is a circle except when $\mu + \nu = 0$, in which case T has the equation $(10x + 12z)z = 0$, which factorizes as the two equations $z = 0$ and $10x + 12z = 0$. Therefore, in the exceptional case, T consists of the line at infinity and the line L with the euqation $5x + 6z = 0$, which is at a right angle to the line joining the centres of the circles and is the locus of points through which the tangents to R and S have the same length. We found L as a line passing through the points of intersection of R and S, but the diagram (figure 2) suggests that the circles R and S do not meet. Indeed, L meets S where $x = -6z/5$ and $y^2 = 2xz - x^2 = -96z^2/25$, that is, they meet in the points with homogeneous Cartesian coordinates $(-6, \pm 4i \sqrt{6}, 5)$, where the y-coordinate is complex. But the degenerate conic T also contains the line at infinity, which meets S where $x^2 + y^2 = 0$, that is, in the points with homogeneous coordinates $(1, \pm i, 0)$. We deduce that the circles R and S meet in four 'points', but all have complex coordinates and two are on the line at infinity.

Many properties of real figures (like the equation of L) can be obtained from points with complex coordinates, so we

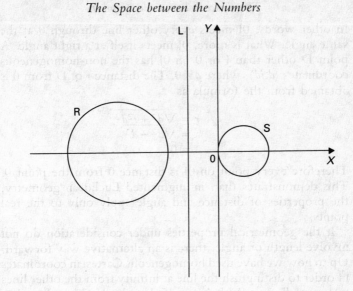

Figure 2 Two circles

allow their use by extending our geometry again to include all the points with complex coordinates, and we have then obtained *augmented Euclidean geometry*. The points I and J with coordinates $(1, i.0)$ and $(1, -i, 0)$ are called the *circular points at infinity* and they have the special property that a (real) proper conic is a circle if and only if it passes through I and J. In fact, the proper conics can be classified in augmented Euclidean geometry by their intersections with the line at infinity.

The use of points with complex coordinates has some important disadvantages. For example, if we use the circular point at infinity J with coordinates $(1, -i, 0)$, we must allow the use of the line OJ, which has the homogeneous equation $y = ix$, the non-homogeneous equation $Y = iX$ and consequently the gradient i. Therefore, according to the standard formula, the tangent t of the angle between OJ and the line through 0 with gradient $g \neq i$ is

$$t = (g - i)/(1 + ig)$$
$$= (g - i)/i(g - i)$$
$$= -i.$$

In other words, 0J meets every other line through 0 at the same angle. What is more, 0J meets itself at a right angle. A point D other than J or 0 on 0J has the non-homogeneous coordinates (d, id), where $d \neq 0$. The distance r of D from 0 is obtained from the formula as

$$r = \sqrt{d^2 + i^2 d^2}$$
$$= \sqrt{d^2 - d^2}$$
$$= 0.$$

Therefore every point on 0J is distance 0 from the point 0. This demonstrates that, in augmented Euclidean geometry, the properties of distance and angle apply only to the real points.

If the geometrical properties under consideration do not involve length or angle, there is an alternative way forward. Up to now we have used homogeneous Cartesian coordinates in order to distinguish the line at infinity from the other lines and to allow calculations of lengths and angles, but abandoning measurement allows the use of more general coordinates. Let (x, y, z) be the homogeneous Cartesian coordinates of a general point in the plane. Then a set of *general homogeneous coordinates* (X, Y, Z) can be obtained by equations

$$X = a_1 x + b_1 y + c_1 z$$
$$Y = a_2 x + b_2 y + c_2 z$$
$$Z = a_3 x + b_3 y + c_3 z,$$

where the coefficients are complex numbers such that these equations can be solved for x, y, z in terms of X, Y, Z. The resulting geometry is called *(Algebraic) Projective Geometry*. In this projective geometry the words 'parallel', 'distance', 'angle' and 'real' have no meaning, but the degree of a curve, cross-ratio, collinearity and concurrency are preserved from augmented Euclidean geometry. Consequently, results like Pappus' Theorem, Desargues's Theorem and Pascal's Mystic Hexagram are true in projective geometry.

It is a very symmetrical geometry, with no parallel lines or real points. There is even symmetry between points and

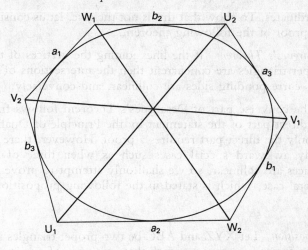

Figure 3   Brianchon's Theorem

lines, the basis of the *Principle of Duality*, which asserts that for every theorem in projective geometry there is a dual theorem in which the words 'point' and 'line' (and the words derived from them) change places. For example, the dual figure of a proper conic is a proper *conic envelope*, the set of tangents to a proper conic. Therefore the dual of Pascal's Mystic Hexagram states that if a_1, a_2, a_3, b_1, b_2, b_3 are tangents to a proper conic S, the line p joins the intersection U_1 of a_2 and b_3 to the intersection U_2 of a_3 with b_2, the line q joins the intersection V_1 of a_3 with b_1 to the intersection V_2 of a_1 with b_3 and the line r joins the intersection W_1 of a_1 with b_2 to the intersection W_2 of a_2 with b_1, then p, q and r are concurrent. With the aid of figure 3, we can restate this as the following.

Brianchon's Theorem   In a plane projective geometry, if the sides of a hexagon are tangents to a proper conic then the diagonals are concurrent.

The absence of length and angle from projective geometry might suggest that proofs must always involve the use of

91

coordinates. To show that this is not the case, let us consider the proof of the following theorem.

Desargues's Theorem   If the lines joining the vertices of two proper triangles are concurrent then the intersections of the corresponding sides are collinear, and conversely.

The converse part of Desargues's Theorem follows from the direct part of the statement by the Principle of Duality, so only the direct part requires a proof. However, there are many awkward special cases such as when three of the vertices are collinear, so we shall only attempt to prove the general case, which is stated in the following proposition.

Proposition   Let XYZ and ABC be two proper triangles in a plane projective geometry such that no vertex of one is on a side of the other and such that the lines AX, BY and CZ are concurrent in a point I. Let D be the intersection of YZ with BC, let E be the intersection of ZX with CA and let F be the intersection of XY with AB. Then the points D, E, and F are collinear.

Proof   We start by defining some more points. The distinct lines BC and ZX meet in a unique point S and the distinct lines ZC and XB meet in a unique point T. Because no vertex of one triangle XYZ or ABC lies on a side of the other triangle, the point I is on neither triangle and the four lines IS, IX, IY and IZ are distinct. Therefore the line XY meets the line IS in a unique point U and the line AB meets the line IS in a unique point V different from U. The diagram for this case of Desargues's Theorem with the additional points S, T, U and V appears as figure 4. We now prove the proposition by three applications of Pappus' Theorem.

First, we apply Pappus' Theorem to the points X, B, T on the line XB and the points V, U, S on the line IS. The reason for this choice is that the intersection of VE = AB with UX = XY is F, and the other two intersections

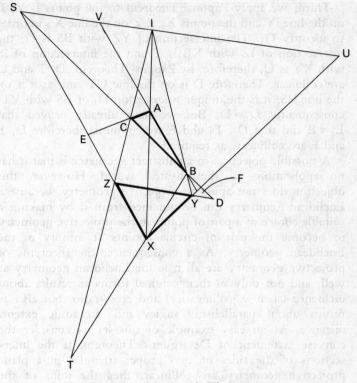

Figure 4 Desargues's Theorem

involved in Pappus' Theorem are on $BS = BC$ and $XS = ZX$.
Let D_1 be the intersection of BS with UT and let E_1 be the
intersection of XS with VT. Then, by Pappus' Theorem,
D_1, E_1 and F are collinear.

Second, we apply Pappus' Theorem to the points I, X, A
on the line IX and B, C, S on the line BS in order to identify
E_1. The intersection of XS with AC is E, the intersection of
IS with AB is V and the intersection of IC with XB is T,
therefore, by Pappus' Theorem, E, V and T are collinear.
Therefore E is on the line VT and, as it is also on the line
XS, it is the unique intersection E_1 of XS with VT,
consequently $E_1 = E$.

Third, we apply Pappus' Theorem to the points I, Y, B on the line IY and the points X, S, Z on the line XS in order to identify D_1. The intersection of YZ with BS is D, the intersection of IZ with XB is T and the intersection of IS with XY is U, therefore, by Pappus' Theorem, D, T and U are collinear. Therefore D is on the line UT and, as it is on the line BS, it is the unique intersection D_1 of BS with VT, consequently $D_1 = D$. But we have already proved that $E_1 = E$ and that D_1, E_1 and F are collinear, therefore D, E and F are collinear, as required.

A possible objection to an abstract geometry is that it has no applications to the physical world. However, this objection does not apply to projective geometry, because a Euclidean geometry can be obtained from it by making a suitable choice of a pair of points in the projective geometry to become the pair of circular points at infinity of the Euclidean geometry. As a consequence, the theorems of projective geometry are all true for Euclidean geometry as well, and not only in their original forms as results about incidence (such as collinearity) and cross-ratio, but also as results about parallelism, angles and, to some extent, distance. As an easy example of this let us consider the convese statement of Desargues's Theorem: if the intersections of the sides of two proper triangles in a plane projective geometry are collinear, then the joins of the corresponding vertices are concurrent. To obtain a theorem of Euclidean geometry, we must choose the circular points at infinity. Let us choose them so that all the lines and points of the figure are real and the line of collinearity is the line at infinity. This implies that, because the corresponding sides of the two triangles meet on the line at infinity, when the joins of the vertices would be parallel. This is a proof of the following.

Theorem If the sides of two proper triangles in plane Euclidean geometry are parallel, then the joins of the corresponding vertices are either concurrent or all parallel.

Solution

Finally, we must use our knowledge of geometry to help the colour-blind bear hunter. She had stalked due west between sightings of the bear, so, unless she was at the equator, she followed an arc of a circle. At the beginning and end of the stalk she saw the place where the bear was feeding at the intersection of two northward lines. That is, at the north pole. Therefore it was a WHITE BEAR.

7

Can you Solve it?: Algebra

'All devotees, resolving,
A pilgrimage must make,
To Rathkeale's *Castle Matrix*,
Before this life forsake.'
Stephen Barnett and Henry Power, 'Castle Matrix'

Algebra is the part of mathematics concerned with the identification of numbers which satisfy conditions expressed in terms of finite mathematical processers, such as addition, but not infinite processes, such as the calculation of limits. This is well expressed in the Arabic phrase from which the word 'algebra' is derived: *al-jabr*, meaning 'the reunion of broken parts'. The word and the subject both reached Europe from the Arabian school of algebra of the early middle ages (from about 750 AD). Algebra was such an important part of Arabian mathematics that it is almost certain that Scheherazade entertained the Sultan by solving algebraic problems and not by telling the stories related in *The Thousand and One Nights*, and the Sultan relented when she pointed out the elegant factorization of 1001 as $7 \times 11 \times 13$.

The easiest part of algebra is *linear algebra*, which studies every aspect of equations of degree 1. As there is no difficulty in solving one linear equation $ax = b$ in one unknown x, the subject starts seriously with a discussion of several simultaneous linear equations in a number of unknowns. The easiest examples soon convince the student that there is, in principle, a straightforward way to solve n equations in n unknowns, for any positive integer n. Consequently, there should be no difficulty in solving two equations in two unknowns. So let us solve the equations

$$x + y = 1,$$
$$x + y = 0.$$

Of course, these have no solution, since otherwise $1 = 0$. But why ever consider such a foolish pair of equations? The answer is that, in Cartesian geometry, these equations represent the lines with gradient -1 through the points with coordinates $(1,0)$ and $(0,0)$, respectively. Because the lines are parallel, we would be horrified if we found a solution for this pair of equations. This brings us to the first problem of linear algebra: which sets of linear equations are *consistent*, that is have at least one solution?

To solve this problem, we need to apply a method of solution which eliminates the danger of finding a solution which does not satisfy all the equations. For example, is there a common point of intersection of the four planes with equations $x + y = 2$, $x - y = 0$, $z = 0$ and $z = 1$ in three-dimensional Cartesian geometry? Because the last pair of planes are parallel, there is no common point, but if we ignore the last equation we can obtain the solution $x = 1$, $y = 1$, $z = 0$. Note that this would have been a correct solution if the fourth plane had been the one with the equation $x + z = 1$.

An alternative problem would be to find a point on the intersection of the planes with equations $x + y = 2$ and $x - y = 0$. Therefore, in order to cover all possible geometrical applications, we must allow the number of equations in our discussions to be independent of the number of unknowns. Consequently, we aim to find a set of equations that can be solved easily and which is *equivalent* to the original set, in that it has the same set of solutions. This set of solutions will be the *empty set*, which contains nothing and is denoted by Ø, whenever the set of equations is inconsistent.

The easiest way is to use the *elementary* operations of exchanging equations, or multiplying an equation by a non-zero number, or adding a multiple of an equation to another, to transform the set of equations into a set in *echelon form*; that is, with each equation having more unknowns with zero coefficients at the beginning of the equation than the previous equation, or containing no unknowns. For example, the set of equations

$$2x + y - 3z \qquad = 1,$$
$$3y + z + \quad w = 2,$$
$$3w = 5,$$
$$0 = 0,$$
$$0 = 0$$

is in echelon form. A set of equations is consistent if and only if an equivalent set in echelon form contains no equation $0 = d$, where d is not 0. For a consistent set of equations in echelon form, the unknowns for which the value may be chosen arbitrarily (such as z in this example) can easily be identified, and the set of solutions can be found systematically. This technique is sufficient for the solution of a wide range of practical problems, such as the following.

Problem

After the unexpected marriage of his daughter to the Crown Prince and the unfortunate death of his second wife, Baron von Haerdupp improved his financial position by selling Schloss Haerdupp and using the proceeds to reduce the mortgage on his country estate, Lerchenau. As a consequence of this change, he reverted to the use of his family name instead of the territorial title 'von Haerdupp'. He calculated that he could soon clear the remaining mortgage with his increased agricultural income if he enclosed his part of the large rectangular field which he shared with the neighbouring landlord, Herr von Sumpfhuhn. The Lerchenau charter stated that Sumpfhuhn's share of the field was from the southern edge up to a line defined in terms of the grazing of geese, goats and horses, but which could be interpreted as indicating a line with an equation $ax + by + c = 0$ with respect to Cartesian axes along the southern and western boundaries of the field, and for which the coefficients satisfy the equation $a + 2b + 3c = 2$. The Sumpfhuhn charter stated the same conditons, but specified the equation $3a + 4b + 5c = 8$. The medieval court records agreed with the charters, but specified the equation $4a + 7b + 10c = 2$. The two landlords accordingly asked the court for a ruling as to where the line should be drawn. The judge decided that all three documents were valid, so the three equations all applied to the determination of the line, but he sent the case to the High Court for a decision. What should the decision of the High Court be?

The application of elementary operations is sufficient for the determination of the consistency and solution of a particular set of linear equations, but we need further methods to discuss a general set of linear equations such as

$$a_1x + a_2y = d,$$
$$b_1x + b_2y = e,$$
$$c_1x + c_2y = f,$$

where a_1, a_2, b_1, b_2, c_1, c_2, d, e and f represent constant real numbers. Let us suppose that these equations represent three lines in a plane Cartesian geometry in which we change the coordinates by rotating the axes. The coordinates (x,y) are related to the new coordinates (X,Y) by equations of the form

$$x = r_1X + s_1Y,$$
$$y = r_2X + s_2Y,$$

where r_1, r_2, s_1 and s_2 are constant real numbers which satisfy certain conditions. The new form of the equations of the three lines is

$$(a_1r_1 + a_2r_2)X + (a_1s_1 + a_2s_2)Y = d,$$
$$(b_1r_1 + b_2r_2)X + (b_1s_1 + b_2s_3)Y = e,$$
$$(c_1r_1 + c_2r_2)X + (c_1s_1 + c_2s_2)Y = f.$$

Because the right-hand sides of these equations are unaltered by the transformations, we shall ignore them and concentrate on the coefficients on the left-hand sides.

To assist in this, we can write the coefficients on the left-hand sides of the original equations in their correct positions in a rectangular array that we call a *matrix*:

$$\mathbf{M} = \begin{bmatrix} a_1 & a_2 \\ b_1 & b_2 \\ c_1 & c_2 \end{bmatrix}.$$

Similarly, the coefficients in the transformation can be written as a matrix

$$\mathbf{T} = \begin{bmatrix} r_1 & s_1 \\ r_2 & s_2 \end{bmatrix}$$

and the coefficients of the left-hand side of the transformed equations as

$$\mathbf{P} = \begin{bmatrix} a_1r_1 + a_2r_2 & a_1s_1 + a_2s_2 \\ b_1r_1 + b_2r_2 & b_1s_1 + b_2s_2 \\ c_1r_1 + c_2r_2 & c_1s_1 + c_2s_2 \end{bmatrix}.$$

We define the matrix $\mathbf{P}$ to be the *product* $\mathbf{MT}$ of the matrices $\mathbf{M}$ and $\mathbf{T}$, because this formula enables us to write the transformed equations immediately, without an intermediate calculation. In fact, the information concerning many mathematical objects can be written as a matrix such that its product by another matrix represents the results of transformation by means of linear equations. The definition of the product of two matrices is easily remembered once its pattern is understood.

Let us look at the product $\mathbf{P} = \mathbf{MT}$:

$$\begin{bmatrix} a_1 & a_2 \\ b_1 & b_2 \\ c_1 & c_2 \end{bmatrix} \begin{bmatrix} r_1 & s_1 \\ r_1 & s_2 \end{bmatrix} = \begin{bmatrix} a_1r_1 + a_2r_2 & a_1s_1 + a_2s_2 \\ b_1r_1 + b_2r_2 & b_1s_1 + b_2s_2 \\ c_1r_1 + c_2r_2 & c_1s_1 + c_2s_2 \end{bmatrix}.$$

Each *row* of **M** has its own letter a or b or c and each *column* of **T** has its own letter r or s. This makes it clear that a row of **P** is constructed from the corresponding row of **M** and that a column of **P** is constructed from the corresponding column of **T**. In particular, the entry in **P** in, say the second row and first column is obtained by reading along the second row of **M** (that is: b_1 b_2) and down the first column of **T** (that is: r_1 r_2), multiplying the corresponding entries (that is: b_1r_1 and b_2r_2) and adding to obtain $b_1r_1 + b_2r_2$. It follows that the product **AB** of two given matrices **A** and **B** can be calculated provided that **A** has the same number of entries in its rows that **B** has in its columns. In other words, **AB** is defined if and only if the number of columns of **A** is equal to the number of rows of **B**.

The number of rows of **A** and the number of columns of **B** do not affect the existence of **AB** but affect the shape of **AB**, because **AB** has the same number of rows as **A** and the same number of columns as **B**. In particular, if **A** and **B** both have n rows and n columns, where n is a positive integer, then the products **AB** and **BA** both exist and both have n rows and n columns. The temptation to assume that **AB** = **BA** must be resisted, however, even in the case $n = 2$. Indeed, if the matrices **A** and **B** are chosen at random, it will probably be found that **AB** $\neq$ **BA**. For example, if

$$\mathbf{AB} = \begin{bmatrix} 0 & 1 \\ 2 & 2 \end{bmatrix} \begin{bmatrix} 1 & -3 \\ 1 & 2 \end{bmatrix} = \begin{bmatrix} 1 & 2 \\ 4 & -2 \end{bmatrix}$$

then

$$\mathbf{BA} = \begin{bmatrix} 1 & -3 \\ 1 & 2 \end{bmatrix} \begin{bmatrix} 0 & 1 \\ 2 & 2 \end{bmatrix} = \begin{bmatrix} -6 & -5 \\ 4 & 5 \end{bmatrix},$$

so $\mathbf{AB} \neq \mathbf{BA}$. This result does not seem so surprising if $\mathbf{AB}$ is thought of as the action of first applying $\mathbf{A}$, then applying $\mathbf{B}$. This action could, for example, be the changing of Cartesian coordinates: if $\mathbf{A}$ is the matrix of the transformation of the first coordinates into the second and $\mathbf{B}$ is the matrix of the transformation of the second coordinates into the third, then $\mathbf{AB}$ is the matrix of the transformation of the first into the third. Let us denote some other activities by letters in a similar way. For example, let P represent 'Put on your parachute' and let J represent 'Jump out of the plane.' The instruction PJ then means 'Put on your parachute and jump out of the plane.' Would you prefer the instruction JP?

It is now time to apply matrix multiplication to linear equations. Consider the matrices

$$\mathbf{C} = \begin{bmatrix} a_1 & a_2 \\ b_1 & b_2 \end{bmatrix}, \quad \mathbf{z} = \begin{bmatrix} x \\ y \end{bmatrix} \text{ and } \mathbf{k} = \begin{bmatrix} d \\ e \end{bmatrix},$$

where a_1, a_2m b_1, b_2, d and e are real numbers and x and y are unknowns.

Then the set of linear equations

$$a_1 x + a_2 y = d,$$
$$b_1 x + b_2 y = e$$

can be written as $\mathbf{Cz} = \mathbf{k}$.

Also, for the matrices

$$\mathbf{T} = \begin{bmatrix} r_1 & s_1 \\ r_2 & s_2 \end{bmatrix} \text{ and } \mathbf{Z} = \begin{bmatrix} X \\ Y \end{bmatrix},$$

where r_1, r_2, s_1 and s_1 are real numbers and X and Y are unknowns, the transformation

$$x = r_1 X + s_1 Y,$$
$$y = r_2 X + s_2 Y$$

can be written as $\mathbf{z} = \mathbf{T} \mathbf{Z}$.

Let us substitute this formula for $\mathbf{z}$ in the equations $\mathbf{C}$ $\mathbf{z} = \mathbf{k}$. Because the associative law (which we discussed in chapter 2) holds for the multiplication of matrices, the linear equations have the form

$$(\mathbf{CT})\mathbf{Z} = \mathbf{k}$$

in the transformed unknowns. Any set of m linear equations in n unknowns, where m and n are positive integers, can be written in the form $\mathbf{Cz} = \mathbf{k}$, so we can find a formula for the solution of the set of linear equations if we can discover a method for dividing matrices. Because $\mathbf{AB} \neq \mathbf{BA}$, a different operation would be needed to divide $\mathbf{AB}$ by $\mathbf{A}$ from that to divide $\mathbf{BA}$ by $\mathbf{A}$. However, for numbers, we can replace the operation 'divide by r' by the operation 'multiply by the reciprocal r^{-1} of r', which allows us to divide d on the left, as $r^{-1}d$, or on the right, as dr^{-1}. In order to adopt this approach for matrices, we first note that, for any positive integer n, there is a matrix $\mathbf{I}$, called the *identity matrix*, such that, for any matrix $\mathbf{A}$ with n rows and columns,

$$\mathbf{AI} = \mathbf{IA} = \mathbf{A}.$$

The entry in the kth row and k the column of $\mathbf{I}$ is 1 for each integer $k = 1, 2, \ldots, n$ and all the other entries in $\mathbf{I}$ are 0. Therefore, for $n = 2$,

$$\mathbf{I} = \begin{bmatrix} 1 & 0 \\ 0 & 1 \end{bmatrix}.$$

The equation $\mathbf{AI} = \mathbf{IA}$ is impossible unless $\mathbf{A}$ is *square*, that ism $\mathbf{A}$ has as many rows a columns. Therefore only a square matrix $\mathbf{A}$ can have the matrix analogue of a reciprocal, an *inverse* matrix, which is a matrix $\mathbf{A}^{-1}$ satisfying

$$\mathbf{A}^{-1}\mathbf{A} = \mathbf{I} \text{ and } \mathbf{AA}^{-1} = \mathbf{I}.$$

If the matrix $\mathbf{C}$ for our set of equations $\mathbf{Cz} = \mathbf{k}$ has an inverse, we can solve the equations by multiplying on the left by $\mathbf{C}^{-1}$ to obtain

$$\mathbf{C}^{-1}(\mathbf{Cz}) = \mathbf{C}^{-1}\mathbf{k}.$$

But $\mathbf{C}^{-1}(\mathbf{Cz}) = (\mathbf{C}^{-1}\mathbf{C})\mathbf{z}$ and $\mathbf{C}^{-1}\mathbf{C} = \mathbf{I}$. Further, $\mathbf{Iz} = \mathbf{z}$, therefore $\mathbf{z} = \mathbf{C}^{-1}\mathbf{k}$.

Not every square matrix, however, has an inverse. Let

$$\mathbf{A} = \begin{bmatrix} 1 & 1 \\ 1 & 1 \end{bmatrix} \text{ and } \mathbf{d} = \begin{bmatrix} 1 \\ 0 \end{bmatrix}.$$

If **A** has an inverse, then the set of linear equations $\mathbf{A}\,\mathbf{z} = \mathbf{d}$ has the solution $\mathbf{A}^{-1}\mathbf{d}$. However, this set of equations, when written out in full, is

$$x + y = 1,$$
$$x + y = 0,$$

which has no solution, as we proved at the beginning of the chapter. On the other hand, it is easy to prove that the matrix

$$\mathbf{B} = \begin{bmatrix} 1 & 3 \\ 0 & 1 \end{bmatrix}$$

has the inverse matrix

$$\mathbf{B}^{-1} = \begin{bmatrix} 1 & -3 \\ 0 & 1 \end{bmatrix}.$$

So, which square matices **C** have inverses? The easiest answer to this question involves the *determinant*, denoted by 'det **C**', which is a number associated with a square matrix. If

$$\mathbf{C} = \begin{bmatrix} a_1 & a_2 \\ b_1 & b_2 \end{bmatrix}$$

then det **C** is defined to be $a_1 b_2 - a_2 b_1$, but the definition is much more complicated for larger square matrices. It can be proved that any square matrix **C** has an inverse $\mathbf{C}^{-1}$ if and only if det $\mathbf{C} \neq 0$. When det $\mathbf{C} \neq 0$, a formula exists for $\mathbf{C}^{-1}$ in terms of determinants, which is useful for proving further results about matrices. In all cases except where **C** has two or three rows, the formula is useless for calculating $\mathbf{C}^{-1}$, but fortunately there is a method of calculation which uses elementary operations and which is easy to use in all cases.

In the case where the set of linear equations $\mathbf{Cz} = \mathbf{k}$ has more than one solution, there is no formula for the solutions, so we ask the alternative question 'How are the different solutions of $\mathbf{Cz} = \mathbf{k}$ related to each other?'. In order to indicate the answer to this question, we need to introduce the sum of two matrices. The matrices $\mathbf{A}$ and $\mathbf{B}$ have a *sum* $\mathbf{A} + \mathbf{B}$ if and only if $\mathbf{A}$ and $\mathbf{B}$ have the same number of rows and columns as each other, and then an entry in $\mathbf{A} + \mathbf{B}$ is defined to be the sum of the corresponding entries of $\mathbf{A}$ and $\mathbf{B}$. In other words, matrices are added componentwise. For example,

$$\begin{bmatrix} 1 & 2 & 2 \\ 7 & 1 & 4 \end{bmatrix} + \begin{bmatrix} 0 & 0 & 1 \\ 4 & -2 & 0 \end{bmatrix} = \begin{bmatrix} 1 & 2 & 3 \\ 11 & -1 & 4 \end{bmatrix}.$$

A matrix $\mathbf{X}$ in which every entry is 0 has the property that, if $\mathbf{C}$ is a matrix with the same number of rows and columns as $\mathbf{X}$, then

$$\mathbf{C} + \mathbf{X} = \mathbf{X} + \mathbf{C} = \mathbf{C}.$$

Therefore $\mathbf{X}$ acts for the addition of matrices in the same way as 0 acts for the addition of numbers, so $\mathbf{X}$ is called a *zero-matrix* and denoted by $\mathbf{O}$. The zero matrix $\mathbf{O}$ with n rows and n columns (where n is a positive integer) also has the property that, if $\mathbf{A}$ has n rows and n columns, then $\mathbf{OA} = \mathbf{AO} = \mathbf{O}$.

Now let us apply this idea to the solution of m equations in n unknowns in the form $\mathbf{Cz} = \mathbf{k}$, where m and n are positive integers, $\mathbf{C}$ is a matrix with m rows and n columns, $\mathbf{z}$ is a *column vector* (that is, a matrix with one column) with n rows each consisting of an unknown and $\mathbf{k}$ is a column vector with m rows. If the unknowns in $\mathbf{z}$ are x_1, x_2, . . . , x_n then the equations $\mathbf{Cz} = \mathbf{k}$ have the solution $x_1 = u_1$, $x_2 = u_2$, . . . , $x_n = u_n$ if and only if $\mathbf{Cu} = \mathbf{k}$, where $\mathbf{u}$ is the column-vector of which the n entries are u_1, u_2, . . . , u_n. Accordingly, we can regard the column-vector $\mathbf{u}$ as a solution of the set of equations $\mathbf{Cz} = \mathbf{k}$. Let the column-vector $\mathbf{v}$ be another solution of $\mathbf{Cz} = \mathbf{k}$. Then $\mathbf{Cu} = \mathbf{Cv} = \mathbf{k}$

and therefore, by the rules of matrix algebra, $C(u - v) = o$, where o is a column-vector with m rows each of which is the number 0. Therefore any solution v of $Cz = k$ is of the form $u + w$, where $Cw = o$. Consequently, we can determine all the solutions of $Cz = k$ by finding just one solution of these equations and then finding all the solutions of the *homogeneous equations* $Cz = o$.

This step simplifies our problem because the solutions of a set of homogeneous equations conform to a regular pattern which is lacking for the non-homogeneous equations. To indicate this, we first introduce the *product of the vector u by the scalar r*, where r is a real number. If

$$u = \begin{bmatrix} u_1 \\ u_2 \\ \cdot \\ \cdot \\ \cdot \\ u_n \end{bmatrix} \quad \text{then } ru = \begin{bmatrix} ru_1 \\ ru_2 \\ \cdot \\ \cdot \\ \cdot \\ ru_n \end{bmatrix}.$$

Now, let V be the set of all solutions of the m homogeneous equations $Cz = o$ in n unknowns, let u and v belong to V and let r be a real number. Then, by the rules of matrix algebra,

$$\begin{aligned} C(u + v) &= Cu + Cv \\ &= o + o \\ &= o, \end{aligned}$$

therefore $u + v$ belongs to V. Further,

$$\begin{aligned} C(ru) &= r(Cu) \\ &= ro \\ &= o, \end{aligned}$$

therefore ru belongs to V as well. This implies that V is an example of a *vector space*. Therefore the properties of the set of solutions of $Cz = o$ can be obtained from results on vector spaces provided the vector space of solutions can be identified.

A vector space V is completely characterized by the set of numbers under consideration (in our case, the set of real numbers) and one integer, the *dimension d* of V, which is the least integer such that there is a set of vectors $\{\mathbf{b}_1, \mathbf{b}_2, \ldots, \mathbf{b}_d\}$ in V such that every vector $\mathbf{v}$ of V can be written as

$$\mathbf{v} = r_1\mathbf{b}_1 + r_2\mathbf{b}_2 + \ldots + r_d\mathbf{b}_d,$$

where $r_1, r_2, \ldots, r_d$ are real numbers. This implies that the question 'What are the relations between the solutions of the equations $\mathbf{Cz} = \mathbf{k}$?' has been reduced to the question 'Can we find the dimension of the vector space of solutions of $\mathbf{Cz} = \mathbf{o}$ in terms of the matrix $\mathbf{C}$?'. The answer to this question is obtained by using another vector space, which consists of all expressions

$$a_1\mathbf{c}_1 + a_2\mathbf{c}_2 + \ldots + a_n\mathbf{c}_n,$$

where $\mathbf{c}_1, \mathbf{c}_2, \ldots, \mathbf{c}_n$ are the columns of the matrix $\mathbf{C}$ and the values of $a_1, a_2, \ldots, a_n$ range over the real numbers. The dimension of this vector space is called the *rank r* of $\mathbf{C}$, and it can be calculated very easily because r is the number of non-zero rows of any matrix $\mathbf{D}$ in echelon form which is equivalent to $\mathbf{C}$. If r is the rank of $\mathbf{C}$ and n is the number of columns of $\mathbf{C}$ (which is equal to the number of unknowns) then the dimension of the vector space V of solutions of $\mathbf{Cz} = \mathbf{o}$ is $n - r$. Consequently, the theory of vector spaces is the key to the properties of linear equations and therefore *linear algebra* can be regarded as the study of vector spaces.

University mathematics syllabuses frequently list separate courses in linear algebra and *abstract algebra* or *Galois theory*. These titles do not indicate the relationship between the courses quite as well as older title *theory of equations*, which reveals that the course studies polynomial equations of degree greater than one. Our account of abstract algebra will be concerned only with the relationships between its principal topics, and this will serve to show that the initial separation of algebra into two parts is only for convenience. We start with the middle ages, when it was known that the general quadratic equation

$$ax^2 + bx + c = 0,$$

where $a \neq 0$, has the solutions

$$x = \frac{-b \pm \sqrt{b^2 - 4ac}}{2a}.$$

The character of these solutions is determined by the *discriminant*

$$\Delta \equiv b^2 - 4ac.$$

If $\Delta = 0$ the solutions are equal, if $\Delta > 0$ there are two real, different solutions, and if $\Delta < 0$ we say that the solutions are complex and different; but, no doubt, the medieval algebraists deduced at first that there were no solutions. Naturally, the medieval algebraists tried to find a similar formula for the solutions and a discriminant for the general cubic equation. This was achieved by Scipio del Ferro, Niccolo Fontana (known as 'Tartaglia') and Girolamo Cardano, who worked in Bologna. They found that it was better to start by transforming the general cubic equation

$$Ax^2 + Bx^2 + Cx + D = 0,$$

where $A \neq 0$, into the *reduced form*

$$x^2 + ax + b = 0.$$

They determined the discriminant of the reduced form as

$$\Delta \equiv \frac{b^2}{4} + \frac{a^3}{27},$$

and Tartaglia determined the general solution of the reduced form by the formula

$$x = \sqrt[3]{-\tfrac{1}{2}b + \sqrt{\Delta}} + \sqrt[3]{-\tfrac{1}{2}b - \sqrt{\Delta}}.$$

This is now known as *Cardano's Formula* because Cardano published it in *Ars Magna* in 1545. The next development in algebra is most easily appreciated by means of an example.

Let us solve the cubic equation (in reduced form)

$$x^3 - 3x + 1 = 0.$$

Can you Solve it?

The discriminant

$$\Delta = -\tfrac{3}{4},$$

so Cardano's Formula gives the general solution

$$x = \sqrt[3]{-\tfrac{1}{2} + \tfrac{1}{2}i\sqrt{3}} + \sqrt[3]{-\tfrac{1}{2} - \tfrac{1}{2}i\sqrt{3}}.$$

By using the properties of an equilateral traingle, which has three angles measuring $\pi/3$ radians, we can write this formula in the neater trigonometrical form

$$x = \sqrt[3]{\cos 2\pi/3 + i \sin 2\pi/3} + \sqrt[3]{\cos 2\pi/3 - i \sin 2\pi/3}.$$

Faced with formulae like these, the medieval algebraists proceeded with trepidation to extract the cube roots of the numbers which they dubbed 'imaginary'. In the eighteenth century, mathematicians like Euler and Abraham de Moivre approached complex numbers more boldly, despite the continued absence of any justification for their use, and proved some surprising results, like the following.

De Moivre's Theorem

Let θ be a real number and let r be a rational number. Then the finite number of values of $(\cos \theta + i \sin \theta)^r$ are given by the different values of

$$\cos r(\theta + 2k\pi) + i \sin r(\theta + 2k\pi),$$

where k ranges over the integers.

If we apply De Moivre's Theorem to our formula, we obtain one solution of $x^3 - 3x + 1 = 0$ as

$$x = (\cos \frac{2\pi}{9} + i \sin\frac{2\pi}{9}) + (\cos \frac{-2\pi}{9} + i \sin\frac{-2\pi}{9})$$

$$= 2 \cos \frac{2\pi}{9},$$

and, similarly, the other two solutions $2 \cos 8\pi/9$ and $2 \cos 14\pi/9$. At this point, medieval mathematicians would (very sensibly) check that these numbers really are solutions of the equation, because they had no confidence in their calculations. We have no such need because the great Irish

109

mathematician Sir William Rowan Hamilton (1805–65) provided a rigorous construction of the complex numbers based on the real numbers. However, we should note in passing that a cubic equation has three real solutions if and only if the discriminant is negative, so the calculations for the solutions necessarily involve the extraction of cube roots of complex numbers.

Cardano's *Ars Magna* also included a method for solving quartic equations (that is, equations of degree 4) which was due to Ludovico Ferrari (1522–65) and which involved the solution of a resolvent cubic equation. In 1675, James Gregory devised a common method for solving both cubic and quartic equations by means of resolvent equations of one degree fewer, and he was applying this method to the general quintic equation (equation of degree 5) when he died later in the year. When other mathematicians continued Gregory's work, they discovered that the resolvent equation is of degree 6, so Gregory's clever method is ineffective for quintic equations.

The formulae for the solutions of quadratic, cubic and quartic equations are in terms of addition, subtraction, multiplication, division and the extraction of roots (such as square roots), and there are called *radical solutions* of the equations. In 1770, Joseph Lagrange, in trying to find radical solutions for polynomial equations of degrees greater than 4, investigated the reasons for the existence of the radical solutions of equations of degrees 2, 3 and 4, and he discovered that the conditions that allowed the methods to work did not hold for equations of any higher degree. From this he could not deduce that quintic equations did not have radical solutions, but only that any radical solutions would be a great deal harder to find.

Lagrange studied the equations by means of the permutations of their solutions. Let the polynomial $f(x)$ of degree n have *roots* $\alpha_1, \alpha_2, \alpha_3, \ldots, \alpha_n$. Then a *permutation* of the roots of $f(x)$ is an operation of rearranging their order of appearance in a list of the roots. (Football pool enthusiasts should beware that they use *combinations*, not permutations.) For example, one permutation p rearranges the roots $\alpha_1, \alpha_2,$

α_3 of a general cubic as α_3, α_1, α_2, and this operation is denoted by

$$p = \begin{pmatrix} \alpha_1\ \alpha_2\ \alpha_3 \\ \alpha_3\ \alpha_1\ \alpha_2 \end{pmatrix}.$$

As the letters α in this symbol for p merely make it less legible, let us omit them and write p as

$$p = \begin{pmatrix} 1\ 2\ 3 \\ 3\ 1\ 2 \end{pmatrix}$$

and regard p as a permutation of 1, 2, 3. If the permutation is followed in action by the permutation

$$q = \begin{pmatrix} 1\ 2\ 3 \\ 2\ 1\ 3 \end{pmatrix}$$

the result is another permutation, which is called the *product*, pq, and it can be calculated as follows.

$$pq = \begin{pmatrix} 1\ 2\ 3 \\ 3\ 1\ 2 \end{pmatrix} \begin{pmatrix} 1\ 2\ 3 \\ 2\ 1\ 3 \end{pmatrix}$$

$$= \begin{pmatrix} 1\ 2\ 3 \\ 3\ 1\ 2 \end{pmatrix} \begin{pmatrix} 3\ 1\ 2 \\ 3\ 2\ 1 \end{pmatrix}$$

$$= \begin{pmatrix} 1\ 2\ 3 \\ 3\ 2\ 1 \end{pmatrix}.$$

In the product qp the operations p and q are performed in the opposite order and

$$qp = \begin{pmatrix} 1\ 2\ 3 \\ 1\ 3\ 2 \end{pmatrix} \neq pq,$$

as might be expected from the similar phenomenon for square matrices.

Because the coefficients of a polynomial $f(x)$ can be expressed in terms of the roots of $f(x)$, Lagrange was able to find a *group* of permutations of the roots of $f(x)$ which leave the coefficients of $f(x)$ unaltered. (In fact, Lagrange did not use the word 'group', which was introduced by Evariste Galois about 60 years later.) For the general polynomial of degree n, the roots are indistinguishably similar, so its group contains all the permutations of 1, 2, 3, . . . , n, and this group is called the *symmetric group of degree n*. The *order* of a group of permutations is the number of distinct permutations it contains, so the order of the symmetric group of degree n is $n!$. In particular, the order of the symmetric group of degree 3 is 6 and the order of the symmetric group of degree 5 is 120.

In 1824, the Norwegian mathematician Niels Abel continued Lagrange's work with the symmetric group of degree 5 and proved that the general quintic equation does not have a radical solution. Note that this states that there is no formula using radicals which solves *all* quintic equations, but it does not imply that *no* quintic equation has a radical solution. Further, it does not imply that there is no method for finding an exact solution of a quintic equation, but it implies that a general solution must use methods which are more advanced than the extraction of roots. In addition, the methods of numerical analysis are available for finding the approximate values of roots of polynomials. Finally, the methods of Abel and Galois have been used to prove that the general polynomial equation of degree n has no radical solution if $n > 4$.

The development that transformed the theory of equations into abstract algebra started with a minor difficulty in the discussion of the group of an equation. For example, the set of six permutations in the symmetric group S_3 for the general cubic equation contains the subset

$$T = \left\{ \begin{pmatrix} 1\,2\,3 \\ 1\,2\,3 \end{pmatrix}, \begin{pmatrix} 1\,2\,3 \\ 2\,3\,1 \end{pmatrix}, \begin{pmatrix} 1\,2\,3 \\ 3\,1\,2 \end{pmatrix} \right\}.$$

The set of permutations T is *closed (under multiplication)* in the sense that if p and q belong to T then pq also belongs to T. But is T the group of permutations of a polynomial? In fact, it is the group of permutations of the polynomial $x^3 - 1$, but no use is made of this polynomial in the discussions of the group S_3. Also, it is still not known whether any subset H which is closed in the group of permutations G of a polynomial $f(x)$ is necessarily the group of permutations of another polynomial. The definition of a group of permutations was therefore (possibly) relaxed slightly to be a set of permutations of a finite set (which may be taken to be 1, 2, 3, . . . , n, for a suitable positive integer n) which is closed under multiplication. Then the group T, which is a subset of S_3, is an example of a *subgroup* of S_3.

Every group G of permutations has at least two (uninteresting) subgroups, the whole group G itself and the subset consisting of the *identity permutation e*, which leaves every symbol unaltered. The identity permutation e has the prperty that, for any permutation p, we have

$$ep = pe = p.$$

The multiplication table for a group (of permutations) is called the *Cayley table* when the elements of the group are written in the same order in the first row and column of the table with e first in both. Because the first row and first column of the table are the same as the headings would be, the headings of the multiplication table are omitted. The entry in the row starting with p and in the column with q at the top is the product pq. Let us construct the Cayley table for the symmetric group S_3 of degree 3, of which the elements are the following permutations:

$$e = \begin{pmatrix} 1\,2\,3 \\ 1\,2\,3 \end{pmatrix}, \qquad a = \begin{pmatrix} 1\,2\,3 \\ 2\,3\,1 \end{pmatrix}, \qquad b = \begin{pmatrix} 1\,2\,3 \\ 3\,1\,2 \end{pmatrix},$$

$$c = \begin{pmatrix} 1\,2\,3 \\ 1\,3\,2 \end{pmatrix}, \qquad d = \begin{pmatrix} 1\,2\,3 \\ 3\,2\,1 \end{pmatrix}, \qquad f = \begin{pmatrix} 1\,2\,3 \\ 2\,1\,3 \end{pmatrix}.$$

We then obtain the Cayley table of S_3, which is given in figure 5, by multiplying all the pairs of these permutations together in both orders. Subgroups of S_3 such as $T = \{e, a, b\}$ and $U = \{e, c\}$ are quite easy to detect by means of the

e	a	b	c	d	f
a	b	e	d	f	c
b	e	a	f	c	d
c	f	d	e	b	a
d	c	f	a	e	b
f	d	c	b	a	e

Figure 5 The Cayley table of S_3

Cayley table and, conversely, the Cayley table is easier to understand if the subgroups are known.

During the 1840s, Augustin Cauchy worked on groups of permutations, looking for alternative methods for calculating the products of the permutations. In this way, he came to study groups of matrices under matrix multiplication and thee led him to a modified definition of a group. Consider the set S of all matrices

$$\mathbf{A}_n = \begin{bmatrix} 1 & n \\ 0 & 1 \end{bmatrix},$$

where n ranges over the positive integers. Then, if m is a positive integer,

$$\mathbf{A}_n\mathbf{A}_m = \mathbf{A}_{n+m}.$$

Therefore S is closed under multiplication, although S does not contain an identity element (which would necessarily be the identity matrix $\mathbf{I}$). Further study of any group G of permutations will reveal that for every element p in G there is an *inverse element* p^{-1} in G such that

$$pp^{-1} = p^{-1}p = e,$$

where e is the identity permutation. This necessarily fails for the set of matrices S because it has no identity element. Consequently, Cauchy defined a group M of matrices as a set of matrices which is closed under multiplication, which contains an identity matrix and which contains an inverse matrix for every matrix in M. Notice that earlier in the chapter we discussed matrices as an item of linear algebra, and here matrices appear just as naturally as an item in abstract algebra.

Let us look at an example of a group of matrices, although here it is more convenient to use unorthodox notation for matrices. The matrices

$$e = \begin{bmatrix} 1 & 0 \\ 0 & 1 \end{bmatrix}, \quad a = \begin{bmatrix} 0 & 1 \\ -1 & -1 \end{bmatrix}, \quad b = \begin{bmatrix} -1 & -1 \\ 1 & 0 \end{bmatrix},$$

$$c = \begin{bmatrix} 0 & 1 \\ 1 & 0 \end{bmatrix}, \quad d = \begin{bmatrix} 1 & 0 \\ -1 & -1 \end{bmatrix}, \quad f = \begin{bmatrix} -1 & -1 \\ 0 & 1 \end{bmatrix}$$

form a group of matrices M in which e is the identity and, for example, $a^{-1} = b$. The Cayley table of M is identical to figure 5. As a consequence, M is *isomorphic* to S_3.

Groups, defined by closure under multiplication and the existence of identities and inverses, were used successfully in geometrical problems from 1850 until 1886 when Heinrich Weber pointed out in his *Lehrbuch der Algebra* that it is also necessary to assume that the associative law holds. The applications of group theory up to that time had all been correct because, by good luck, the associative law happened to be valid in all cases considered. A typical use of group theory is to attempt a characterization of a figure by means of its *symmetries*, that is, the movements of the figure which leave it apparently (or, indeed, actually) unchanged. The product of two symmetries s and t is the symmetry st obtained by operating first with s and then with t, that is, st is obtained by *composition* of the mappings s and t. With this

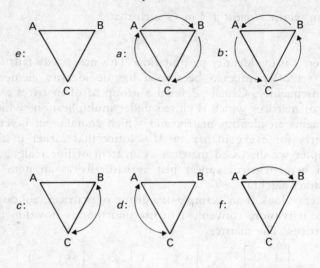

Figure 6 The symmetries of an equilateral triangle

product, the set of symmetries of a geometrical figure (either in the plane or in space) is a group, of which the group T of symmetries of an equilateral triangle ABC, as illustrated in figure 6, is an easy example. The Cayley table of T is identical to figure 5, so T is isomorphic to the group S_3. Similarly, the group U of symmetries of a square is easily seen to contain eight elements. Therefore T and U are not isomorphic and consequently equilateral triangles and squares are distinct figures.

Of course, this is obvious, but whether two complicated crystals are really the same is not so obvious, and this technique can be applied profitably to them. In fact, most applications of group theory in both theoretical and practical problems are related to this idea, consequently groups appear in many guises (such as sets of functions or symmetries) other than sets of permutations or matrices. This is why groups are defined abstractly, as follows. A *group* is a set G on which is defined a *binary operation* (addition, multiplication, composition, uglification, etc.) which (in order to be definite) we name and write as *multiplication* such that:

116

1   G is *closed* under multiplication, that is, for all x and y in G, the product xy belongs to G.

2   Multiplication in G is *associative*, that is, for all x, y and z in G, we have $(xy)z = x(yz)$.

3   G has an *identity* e, that is, for all x in G, we have $xe = ex = x$.

4   For each x in G there exists an *inverse* x^{-1} in G such that $xx^{-1} = x^{-1}x = e$.

The generality of this definition suggests that it includes a far greater number of groups than Lagrange's definition of groups of permutations, by including abstract groups defined by suitable Cayley tables. However, this is denied by the following, which was proved by Arthur Cayley in 1854.

Cayley's Theorem   Every group is isomorphic to a group of permutations.

In working with abstract groups, one has the advantage of meeting any accusation of ignorance with 'Of course I don't know what I'm talking about! But neither do you!'. Another advantage, which is more often employed, is to perform calculations for a group G in the most convenient *faithful representation* of G, that is, a group of matrices or permutations (or whatever) isomorphic to G. This indicates that abstract group theory is an example of *soft mathematics*, because a problem can be modified by considering a different representation or by expressing the conditions of the problem in a different form. This contrasts with *hard mathematics* in which the solution of each problem is required in specific terms, such as 'Find a radical solution of $x^5 - 6x + 3 = 0$.' (This is not only hard but also impossible.) However, it must be realized that hard mathematics is often easy and soft mathematics is often hard.

Solution

By a happy coincidence, when the High Court of Ruritania met to decide on the boundary between Lerchenau and

Can you Solve it?

Sumpfhuhn, Arthur Cayley happened to be touring the country with the Alpine Club. (This was probably the visit on which he painted the view of Lerchenau which is now in the National Gallery of Ruritania.) Because Cayley was a lawyer as well as a mathematician, he was invited to act as a mathematical assessor to the Court. He accepted the post, and started with the equations

$$a + 2b + 3c = 2,$$
$$3a + 4b + 5c = 8,$$
$$4a + 7b + 10 = 2.$$

He subtracted three times the first equation from the second and four times the first equation from the third to produce the following set of equations, which is equivalent to (that is, has the same solutions as) the first set:

$$a + 2b + 3c = 2,$$
$$-2b - 4c = 2,$$
$$-b - 2c = -6.$$

Then he subtracted half the second from the third to give the equivalent set:

$$a + 2b + 3c = 2,$$
$$-2b - 4c = 2,$$
$$0 = -7.$$

Cayley deduced that the equations have no solution from the fact that $0 = -7$ has none, so the High Court awarded the entire field to Herr von Sumpfhuhn. This serious loss led Baron von Haerdupp (or Baron Ochs, as he was known at that time) to consider improving his financial position by marrying into a rich mercantile family that he had met in Vienna. After consulting his cousin, the wife of Field Marshall von Werdenburg, he settled the matter by sending the symbol of betrothal, a silver rose, to his fiancée in the care of his distant cousin, Count Oktavian Rofrano.

8

Little by Little:
Calculus

'The rate of inflation is increasing more slowly.'

Newspaper headline

When looking through the newspaper for the sports page, I frequently see headlines concerning inflation. What is inflation and why should it matter to anyone whose tyres are not flat? In fact, the headlines refer to *financial* inflation, which is the cause, not the effect, of a lot of hot air. The fuel that provides the heat is an *index*, which measures the cost of living by sampling the costs of a balanced selection of goods and services. As the index is calculated only once a month, a graph of the index i against the time t looks something like the graph in figure 7(a). The 'steps' of this graph are due entirely to the fact that the index is calculated at fixed intervals, whereas the cost of living changes frequently by many small amounts. Consequently, a better approximate representation of the change of the cost of living c is the smooth graph in figure 7(b). The graph of the index and the smooth graph of the cost of living both illustrate *inflation*, the decreasing purchasing power of a fixed sum of money.

As inflation is definitely undesirable, the controllers of national finance need to be able to quantify it in order to determine the scale of the action to be taken. Accordingly, the *rate of inflation* is measured as the slope of the (smooth) graph of the cost of living plotted against time, as in figure

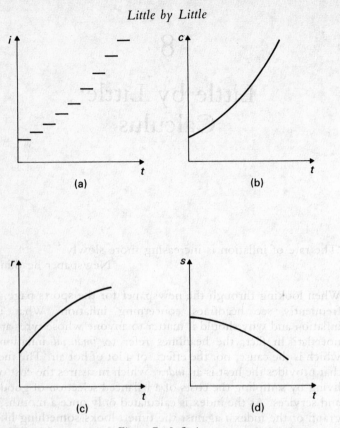

Figure 7 Inflation

7(b). (Economists do not all agree on this definition, or, indeed, on anything else.) The rate of inflation r as obtained from figure 7(b) is plotted against time in figure 7(c), which reveals that the rate of inflation is increasing in this case. A constant rate of inflation means that each month one must make a domestic economy of equal value to last month's, but an increasing rate of inflation implies that successive economies must be ever more stringent. An increasing rate of inflation is therefore very bad news.

Is there no hope? Let us draw another graph in which the slope s of the graph of the rate of inflation is plotted against time. This graph appears as figure 7(d), and it shows a falling curve. Here, at last, is a glimmer of good news: 'The rate of inflation is increasing more slowly.' But what would happen if this downward curve of s against t (the time) continues? Can we use this extended graph to determine the variation of the cost of living? Alternatively, if we know the graph for the cost of living, can we calculate the rate of inflation r and its variation s? These questions are all included in one, more general, question: 'Can we obtain the relation between a graph and its slope by means of calculations?'.

However, before we discuss the slope of a graph we need to agree on what a graph really is. For example, which of the diagrams in figure 8 is a graph?

It is easy to agree that figure 8(a) is a graph, but it is vital to realize that it is not a typical graph but a *smooth* graph, with a steadily changing slope. Figure 8(b) is also a smooth graph, even though all the values of y are negative and there are no points corresponding to positive values of x. The circle in figure 8(c) looks like a smooth graph, but it is not a graph at all because, for example, there are two points associated with $x = 0$ instead of one or none. In figure 8(d) there is supposed to be a point on the left-hand end of each 'step' but not on the right-hand end. Because of this, the figure is a graph, but the gaps imply that it is not *continuous*. Figure 8(e) is a continuous graph with slope -1 for negative x and slope 1 for positive x, but the slope cannot be defined at all for $x = 0$. Consequently, figure 8(e) is a continuous graph which is not smooth. The sawtooth graph of figure 8(f) makes it clear that a continuous graph can have many points where it is not smooth. The dots in figure 8(g) represent points, but they could as easily represent the fruit in a slice of Dundee cake, with two pieces of fruit frequently occurring on the same vertical, in accordance with that splendid traditional recipe, so figure 8(g) is certainly not a graph. The points in figure 8(h) are far more diffuse and may represent a slice of something baked to a recipe in which the fruit is incidental, such as a Yorkshire tea-cake. The scattered points in figure 8(h) form a highly discontinuous graph. The graph in figure 8(i) is formed from figure

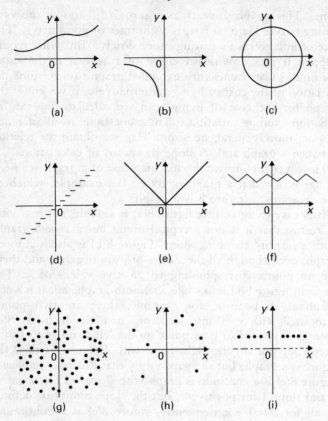

Figure 8 Graphs, real and impossible

8(h) by plotting $(x,1)$ if there is a point (x,y) on the tea-cake graph and plotting $(x,0)$ otherwise. In this case, the x-axis is part of the graph, except for the gaps which occur below the points on the line $y=1$. The function that we used to construct figure 8(i) is called the *characteristic function* of the tea-cake graph. Characteristic functions are difficult to treat mathematically, but they are of considerable importance in statistics and physics.

We may summarize our observations by defining a *graph*,

with respect to given (x, y)-axes, as any figure in which there is at most one point of the figure above, below or on each point of the x-axis. From figure 8 we learn that a graph need not be continuous and continuous graphs need not be smooth.

Let us move the emphasis away from graphs by discussing functions. A *function* is a rule f which defines at most one real number $y = f(x)$ for each real number x. Every graph defines a function in the following way. For each value of x such that there is a point (x, y) on the graph, we define $y = f(x)$, and for every other value of x, we do not define a value of $f(x)$. In this way, figure 8(h) defines a tea-cake function, figure 8(i) defines its characteristic function and figure 8(f) defines a function which is continuous (a term we shall define formally in chapter 9). In these cases, there are no formulae for the functions, but the continuous function of figure 8(e) and the discontinuous function of figure 8(d) and the discontinuous function of figure 8(d) are determined by formulae. Figure 8(e) represents the *absolute value* or *modulus* of x, which is denoted by $|x|$ and defined by $|x| = x$ if $x \geq 0$ and $|x| = -x$ if $x < 0$. Figure 8)d) represents the *integer part* of x, which is denoted by $[x]$ and is defined as the greatest integer which is less than or equal to x. Whenever we have a function, whether it is defined by a formula or a procedure or in some other way (such as by data from an experiment), it is always possible to define its graph, through some graphs (for example, some characteristic graphs) are imposs-ible to draw. However, calculus is concerned with discussing functions with smooth graphs, like those in figures 8(a) and 8(b).

We have now clarified our principal problem concerning graphs: how can we calculate the slope of the smooth graph of a (suitable) function $y = f(x)$? The slope of the graph at a point A is the same as the slope of the tangent at A, which is represented by the line AT in figure 9. The line AT is a tangent to the graph at the point A if the x-component a of A is a double root of the equation in x determining the intersection of AT and the graph. This allows the slope of AT to be found by algebraic methods in a few cases, but this

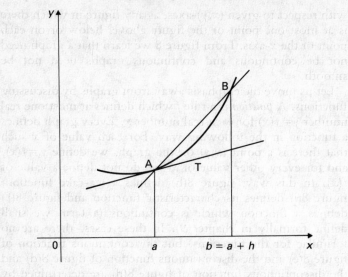

Figure 9 The slope of a graph

method is not helpful in the general case. Instead, we can regard AT as the extreme case of a chord AB of the graph when B coincides with A. Therefore the slope *g* of the chord AB can be regarded as an approximation to the slope of AT, so let us calculate *g*.

Let A and B have the *x*-coordinates *a* and *b*. Because A and B are both on the graph of the function $y = f(x)$, the *y*-coordinates of A and B are $f(a)$ and $f(b)$. The *gradient* of the line segment AB is the difference between the *y*-coordinates of B and A divided by the difference between the *x*-coordinate, that is,

$$g = \frac{f(b) - f(a)}{b - a}.$$

This value *g* might be a sufficiently good approximation to the slope of AT for some practical purposes, but we are treating the problem as pure mathematics, so we must give either an exact answer or a method of obtaining an approximation which is as good as that required by any particular application. In other words, we must treat the problem in a completely general way, in order to cover all applications.

In order to look for an exact answer, let us examine the slope of a variable chord AB, for which purpose we write $b = a + h$, where h is a variable. Notice that h is allowed negative as well as positive values in order to include the chords for which B is to the left of A. By substituting $b = a + h$ in the formula for g, we obtain the slope of the variable chord AB as the value of the formula

$$N(h) = \frac{f(a + h) - f(a)}{h} \, .$$

This formula is called the *Newton quotient* for f at a. We cannot obtain the slope of AT immediately by letting B coincide with A, that is, by putting $h = 0$, because the Newton quotient $N(0)$ is the entirely meaningless formula nothing divided by nothing. This means that a new process, which is called the *limit*, is needed in order to calculate the exact slope of AT.

The notation for the limit of a function $g(x)$ as x tends to the value a is

$$\lim_{x \to a} g(x) = L$$

or

$$g(x) \to L \text{ as } x \to a.$$

I shall delay my discussion of the definition of the limit of a function until the next chapter, but here we should note two facts: in discussing the limit of $g(x)$ as $x \to a$ we insist that x is never equal to a, and, second, the limit of a function does not necessarily exist. We conclude that the *slope* or *gradient* of the graph at A, that is, the slope of AT, is given by

$$\lim_{h \to 0} N(h).$$

This is the limit of the slope of the chord AB as the point B approaches A. We generalize this idea by applying it to the function $f(x)$ instead of the graph of $y = f(x)$. We then say that the *differential coefficient* $f'(a)$ of $f(x)$ at a, if it exists, is given by

$$f'(a) = \lim_{h \to 0} \frac{f(a+h) - f(a)}{h}.$$

(Remember, the limit need not exist, so $f'(a)$ need not exist.) This is essentially Newton's notation for the differential coefficient, with the differential coefficient with respect to x denoted by a dash, and the differential coefficient with respect to t (representing the time, for example), if that is the variable being used, denoted by a dot.

Leibniz's notation for the same idea is also in common use, but it requires some explanation. Leibniz regarded h as 'a small change in the value of x' and he denoted it by δx. As the numerator in the Newton quotient, $f(a+h) - f(a)$, is the corresponding change in $y = f(x)$ and, for the limit to exist, it is small when δx is small, Leibniz wrote

$$\delta y = f(a+h) - f(a).$$

Then, the Newton quotient for f at a is $\delta y / \delta x$ and the differential coefficient is the limit of $\delta y / \delta x$ as $\delta x \to 0$. Consequently, Leibniz wrote the differential coefficient as

$$\frac{dy}{dx} = \lim_{\delta x \to 0} \frac{\delta y}{\delta x}.$$

We conclude that, for a given function $y = f(x)$, the notations $f'(a)$ and dy/dx both represent the same thing, the differential coefficient.

The most important, as well as the first, use of the differential coefficient was to define velocity in dynamics. Let us suppose that we have drawn the graph of the directed (that is, positive or negative) distance s metres of of a particle from the origin along a straight line at time t, measured in seconds. What is the velocity of the particle when its position is represented by a point A at time $t = a$? The slope of a chord AB to the graph is obtained by dividing the distance between A and B by the time taken from $t = a$ to

$t = b$. That is, the slope of the chord AB is the average velocity for the period of $b - a$ seconds after $t = a$. By the argument that we used for the gradient of a graph, the velocity of the particle at time $t = a$ is the slope of the tangent to the graph at A. Therefore the *velocity* of the particle at time $t = a$ is the differential coefficients ds/dt or $\dot{s}(a)$, measured in metres per second. However, we really need to know the velocity of the particle throughout the motion, not merely at a fixed time, and so we require the value of $\dot{s}(a)$ for all relevant values a of t. This is typical of many practical problems, so we give the resulting function a name. For a function $y = f(x)$, the *derived function* is the function, denoted by $f'(x)$ or dy/dx, for which the value at every $x = a$ is the differential coefficient $f'(a)$ of $f(x)$, provided that it exists. The concept of the derived function $f(x)$ opens up a new possibility: we can calculate the derived function of the derived function, to obtain the *second derived function*, which is denoted by $f''(x)$ or d^2y/dx^2. Indeed, we can continue this process and, for a positive integer n, we can define the *nth derived function*, which is denoted by $f^{(n)}(x)$ or $d^n y/dx^n$. In the important case of the motion of a particle in a straight line, the second derived function, $d^2 s/dt^2$ or $\ddot{s}$, is the *acceleration* of the particle. For the example of inflation and the cost of living, which we discussed earlier, we now see that the rate of inflation r is the derived function dc/dt of the (smooth) cost of living c with respect to the time t, and the newspaper headline tells us that the derived function dr/dt (which is also $d^2 c/dt^2$) is decreasing. These examples help us to reformulate the principal problem concerning graphs: we need to calculate the derived function of a given function which has a smooth graph, preferably without any further calculation of limits. The subject in which this problem is studied is *differential calculus*.

The methods of differential calculus assume that the derived function of every 'basic' function is already known. For example, the formula $dy/dx = ax^{a-1}$ when $y = x^a$, for any real number a, is to be remembered. When a new 'basic' function, such as $y = \sin x$, is studied, it is necessary to use a limit process in order to find that $dy/dx = \cos x$, but then this formula is added to the list of formulae to be remembered. Then the derived functions of functions like $x^3 + 2x + 5$, $\sqrt{x^2 + 1}$ and x sin 3x are obtained by means of formulae which determine the derived functions for addition, multiplication, division and composition of given functions. Let $u(x)$ and $v(x)$ be *differentiable* functions of x, that is, functions which have the derived functions du/dx and dv/dx. Then the derived functions of $u + v$, uv and u/v are given by the formulae:

$$\frac{d}{dx}(u+v) = \frac{du}{dx} + \frac{dv}{dx},$$

$$\frac{d}{dx}(uv) = v\frac{du}{dx} + u\frac{dv}{dx}$$

and

$$\frac{d}{dx}\left(\frac{u}{v}\right) = \frac{v\dfrac{du}{dx} - u\dfrac{dv}{dx}}{v^2}.$$

The composite $w(x)$ of $u(x)$ and $v(x)$ is the function in which $v(x)$ is substituted for x in $u(x)$, that is, $w(x) = u(v(x))$. For example, if $u(x) = x^2$ and $v(x) = x^2 + 1$, then $w(x) = (x^2 + 1)^2$. The derived function of the composite is given by the remarkably simple formula called the *Chain Rule*:

$$\frac{dw}{dx} = \frac{du}{dv}\frac{dv}{dx}.$$

The name is based on the chain of derived functions which appears when the Chain Rule is applied successively to the composite of several functions.

As an illustration, let us find the derived function of

$$y = \sin\left(\frac{x^2}{x+1}\right).$$

First, we note that y is the composite $w(x)$ of $u(x) = \sin x$ and $v(x) = x^2/x + 1$. Therefore $w = \sin v$ and, by the formula for the derived function of $\sin x$,

$$\frac{du}{dv} = \cos v.$$

Therefore, by the Chain Rule,

$$\frac{dv}{dx} = \frac{dw}{dx} = \frac{du}{dv}\frac{dv}{dx}$$

$$= \cos v \frac{dv}{dx}$$

$$= \cos\left(\frac{x^2}{x+1}\right)\frac{dv}{dx}.$$

This reduces our problem to the *differentiation*, that is, finding the derived function of,

$$v = \frac{x^2}{x+1}.$$

But

$$v = \frac{u^\star}{v^\star}$$

where $u^\star = x^2$ and $v^\star = x + 1$. By the formula for the derived function of a sum,

$$\frac{d}{dx}(x+1) = \frac{d}{dx}x + \frac{dd}{dx}1.$$

From the formula '$d/dx\, x^a = a\, x^{a-1}$', we obtain

$$\frac{d}{dx}x^2 = 2x, \quad \frac{d}{dx}x = 1 \quad \text{and} \quad \frac{d}{dx}1 = 0,$$

therefore

$$\frac{du^\star}{dx} = 2x \quad \text{and} \quad \frac{dv^\star}{dx} = 1.$$

By the formula for the derived function of a quotient,

$$\frac{dw}{dx} = \frac{v\,\dfrac{du}{dx} - u\,\dfrac{du}{dx}}{v^2}$$

therefore

$$\frac{dv}{dx} = \frac{(x+1)2x - x^2 1}{(x+1)^2}$$

$$= \frac{x^2 + 2x}{(x+1)^2}.$$

We conclude that

$$\frac{dv}{dx} = \frac{x^2 + 2x}{(x+1)^2} \cos\left(\frac{x^2}{x+1}\right).$$

Problem

Those who are skilled at differentiation may like to test their technique by differentiating the following formula, where [x] represents the greatest integer which does not exceed x and ln x represents the natural logarithm of x (for which $d/dx \ln x = 1/x$):

$$y = x^{[\ln \ln \sin^2(x+1)]}.$$

This differentiation problem is unusual in that it is more readily solved by someone of contemplative rather than active disposition.

Usually, differentiation is a straightforward process in which the formulae are employed methodically. However, it is not a process that can be carried out by deploying conventional wisdom, as was pointed out in this typical poem by that prolific poet, Liam-Eric Reiter.

> By moonlight an owl can discriminate
> A mouse from a leaf, but appreciate
> That on darker nights,
> On its hunting flights,
> It can see, but it can't differentiate.

Now let us continue our discussion of calculus by turning to the other major part, called *integral calculus*. The importance of integral calculus can easily be appreciated from its application to one of the most important contemporary problems, the determination of the volume of a hamburger. Of course, it is easy to find the volume of one of those thin, mass-produced hamburgers. The difficulty is to find the volume of a true, wild hamburger, which is flat-bottomed due to its natural affinity with a frying pan but which has a symmetrically-curved upper surface. The method of Archimedes for measuring the volume is to lower the hamburger into a jug of water and measure the water which spills out. This method is unsatisfactory due to its deleterious effect on the flavour of the hamburger, which is consistent with Archimedes' well known indifference to the quality of hamburgers. An ingenious variation of this method is to put the hamburger in a jug of whisky and count the number of Scotsmen needed to drink drams of the overflow. This is *Macdonald's method*, which induces a rich, peaty flavour in the hamburgers and a staggering increase in the population of the kitchen. Nevertheless, a more theoretical approach is to be preferred.

By means of formulae concerning volumes of figures with circular symmetry, the volume of the hamburger can be found provided that we can determine the area of a vertical slice cut through the centre. (This reduction of the problem is not entirely necessary because volumes can be calculated by the same process as areas, but in a more complicated form.) In turn, the area of this slice can be found provided that we can calculate the area of any part of the slice between two vertical cuts. Such a figure can be represented with respect to a set of coordinate axes as the region R bounded by the x-axis, two lines parallel to the y-axis and a part of the smooth graph of a function $f(x)$ which is entirely above the x-axis.

How can the area A of the region R be calculated? According to our original definition, only a rectangle has an area, defined as the product of the length by the breadth. By results in Euclidean geometry, an area can be ascribed to a

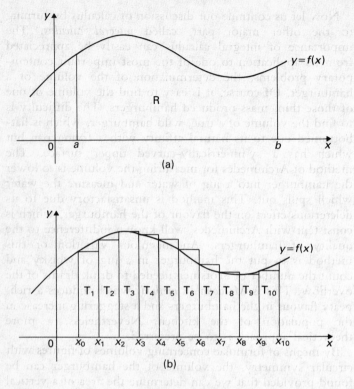

Figure 10 The area under a graph

triangle, equal to half the length of the base multiplied by the
height. As any polygon can be decomposed into a finite
number of triangles, the areas of polygons can be calculated
by elementary methods, but all other areas can only be
calculated by the use of some form of integral calculus.

The area A we wish to define is called the *area under the
curve* $y = f(x)$. It is the area of the region R in figure 10(a),
bounded by the x-axis, the line $x = a$, the line $x = b$ and the
smooth graph of the function $f(x)$. However, the definition
of the area A is so complex that the plot of *War and Peace*
seems simple by comparison. Consequently, we shall give
only a rough indication of the definition.

Our aim is to give a rough definition for the area A of the region R in figure 10(a). We start by choosing a positive integer n and then splitting the part of the x-axis from a to b into n equal segments by means of the points x_0, x_1, x_2, ..., x_j, ..., x_n such that $x_0 = a$, $x_n = b$,

$$x_0 < x_1 < x_2 < \ldots < x_j < \ldots < x_n$$

and $x_j - x_{j-1} = (b-a)/n$, for $j = 1, 2, 3, \ldots, n$. Let us write δx for the length of each of these segments, that is,

$$\delta x = (b-a)/n.$$

Next, for each $j = 1, 2, 3, \ldots, n$, we draw a segment of a vertical line from the point x_{j-1} on the x-axis to the curve $y = f(x)$. The length of this line segment is therefore $f(x_{j-1})$ and the small rectangle T_j of which two sides are this segment and the segment of the x-axis from x_{j-1} to x_j has area $f(x_{j-1})\, \delta x$. The area of T_j is approximately equal to what we think should be the area under curve $y = f(x)$ cut off by the sides of T_j when suitably extended. Let S be the sum of the areas of the small rectangles T_j for $j = 1, 2, 3, \ldots, n$. Then

$$S = \sum_{j=1}^{n} f(x_{j-1})\, \delta x.$$

Figure 10(b) illustrates (with $n = 10$) the fact that S is an approximation to the area A of the region R. Further, if a smaller value of δx is taken (or, equivalently, a larger value of n is taken), the area of the small rectangle T_j is a better approximation to the area under the curve $y = f(x)$ enclosed by the extended sides of T_j, and consequently S is a better approximation to the required value A. We cannot give δx the value 0 because that contradicts the fact that $\delta x = (b-a)/n$, where n is a positive integer. As with the problem of defining the slope of a graph, the way out of this difficulty is to take the limit as δx tends to 0. Therefore we define the *area A under the curve $y = f(x)$ to be*

$$A = \lim_{\delta x \to 0} \sum_{j=1}^{n} f(x_{j-1}) \; \delta x,$$

provided the limit exists. This limit is quite a complicated one, but its expression is more complicated than necessary because n and δx are related quantities. Therefore we simplify the formula to give a still rougher definition of the area by replacing x_{j-1} by the 'typical' value x. Then

$$S = \sum f(x) \; \delta x,$$

where the summation is taken over the values of x which are determined by δx via the formula $\delta x = (b - a)/n$. Then we indicate the area R of the region R by

$$A = \lim_{\delta x \to 0} \sum_{j=1}^{n} f(x_{j-1}) \; \delta x,$$

If we remove the restriction that the function $f(x)$ must be positive between $x = a$ and $x = b$, we no longer obtain the area between the curve $y = f(x)$ and the x-axis because the product $f(x) \; \delta x$ is -1 times the area of the small rectangle. however, this generalization gives us the *integral of the function $f(x)$ between $x = a$ and $x = b$* instead, and this is denoted by

$$\int_{a}^{b} f(x) \; dx = \lim_{\delta x \to 0} \sum f(x) \; \delta x.$$

We conclude that the integral of $f(x)$ between $x = a$ and $x = b$, provided it exists, has a precise definition related to this rough formula.

In books on other subjects, such as physics, integrals often find their way into the calculations by means of this rough definition. This rough definition also helps to explain the strange notation for the integral. The sign $\int$ is really a capital 'long s', the alternative form of a s which waſ ſo often uſed in ſeventeenth century typeſetting. The right-hand side of the definition has Σ (that is, a Greek capital s) to represent a sum, so the left-hand side has $\int$ to indicate a sum which has been altered by the limit. Because $\delta x \to 0$, the symbol δx on the right-hand side is replaced on the left-hand side by its ghost *dx*, just as in the notation *dy/dx* for the derived function. In fact, a notation something like *If*(*x*) or *I*[*a,b*]*f*(*x*) would be more suitable for the integral of *f*(*x*), but the strange notation *dx* has two good features: it is a reminder that *x* is the variable being used (some other variables might appear in the formula for *f*(*x*)) and it simplifies the statement of the Substitution Rule, which we shall consider later.

Our discussion of the definition of the integral indicates the importance of the integral whenever an area is needed, and also whenever a quantity which varies from point to point needs to be summed, but it gives no hint as to how it can be evaluated other than by calculating a complicated limit. In fact, it is relatively easy to obtain an approximate numerical value for an integral by using a precise form of this limit formula and numerical techniques. Computers evaluate integrals in this way. However, integral calculus is devoted to the evaluation of integrals by methods which make only rare use of limits. The key to the subject is the following remarkable and exceedingly important theorem, which was proved independently by Leibniz and Newton.

The Fundamental Theorem of Calculus Let *f*(*x*) be a function with a smooth graph between *x* = *a* and *x* = *b* inclusive, and let *F*(*x*) be a function such that *dF/dx* = *f*(*x*). Then

$$\int_a^b f(x) \; dx = F(b) - F(a).$$

The Fundamental Theorem of Calculus calls for a method of finding a function *F*(*x*) of which the derived function *dF/dx* is a given function *f*(*x*). The function *F*(*x*) is called the *indefinite integral* of *f*(*x*), and we write

$$F(x) = \int f(x) \; dx.$$

This notation is obtained in an obvious way from the notation for the integral, but a notation like $D^{-1}f(x)$ would be more appropriate, where *D* represents the operation of differentiation. However, the indefinite integral is not uniquely defined, because $d/dx\{F(x) + g(x)\} = dF/dx$ whenever $g(x)$ is a function such that $dg/dx = 0$. If $g(x)$ is any constant function, we know that $dg/dx = 0$, and it is also true that if $dg/dx = 0$ then g is a constant function. We therefore cover all the possibilities for an indefinite integral by adding an unknown constant to the result. This constant is not restricted in any way, and is therefore called the *arbitrary constant*. (It is important not to confuse it with the other well-known variable constant, *Cook's constant*, which is the difference between the answer that is required and the answer that is obtained.)

As finding the indefinite integral, or *integration*, is the reverse process of differentiation, we can obtain rules for integration from the rules for differentiation. The rule for differentiating a sum leads immediately to the formula

$$\int \{f(x) + g(x)\} \, dx = \int f(x) \, dx + \int g(x) \, dx.$$

With a little adjustment, the rule for differentiating a product leads to the formula for *integration by parts*:

$$\int f(x)g(x) \, dx = f(x) \int g(x) \, dx - \int \{f'(x) \int g(x) \, dx\} \, dx.$$

This formula is often useful, but is is certainly not a formula for integrating a product of functions in terms of the integrals of the factors. The formula for the derived function of a quotient does not lead to a workable formula for integration, but the Chain Rule leads to the important *Substitution Rule*: if x is a differentiable function of the real variable t, then

$$\int f(x) \, dx = \int f(x(t))\frac{dx}{dt} \, dt.$$

There is an important difference between the use of the Chain Rule for differentiation and the use of the Substitution Rule for integration: the functions for the Chain Rule are obtained by analysing the given function, but the function x of t in the Substitution Rule can be freely chosen according to the integrator's imaginative appreciation of the resulting indefinite integral. Whereas differentiation can be done systematically by the formulae, integration requires ingenuity and intuition. Furthermore, some indefinite integrals cannot be evaluated at all in terms of the functions already in use, but they can be used to define new functions. For example, it is easy to prove that

$$\int x^a \, dx = \frac{x^{a+1}}{a+1} + A,$$

where A is a constant, provided that $a \neq -1$, but the case $a = -1$ can be used to define the *natural or Napierian logarithm*

$$\ln x = \int_1^x \frac{dt}{t}.$$

Solution

By now the fruits of contemplation of the formula

$$y = x^{[\ln \ln \sin^2(x + 1)]}$$

should be ripe, so let us gather the harvest. Because $[x]$ is an integer, the derived function of y should be easy to calculate once we have evaluated $[\ln \ln \sin^2(x + 1)]$. Whatever value $x + 1$ takes,

$$-1 \leq \sin (x + 1) \leq 1$$

therefore

$$0 \leq \sin^2(x + 1) \leq 1.$$

Whenever $\sin^2(x + 1) = 0$, the function $\ln \sin^2(x + 1)$ is not defined, but for the other values, $\ln \sin^2(x + 1)$ is defined but negative. Consequently, $\ln \ln \sin^2(x + 1)$ is not defined as a real function.

Those who know that it does have an existence as a complex function will then wish to evaluate

$$[\ln \ln \sin^2(x + 1)],$$

which is the greatest integer less than or equal to $\ln \ln \sin^2(x + 1)$, a concept which is certainly not defined, because 'less than' is meaningless for complex numbers. Therefore [ln ln $\sin^2(x + 1)$] is not defined and, a fortiori (that is, deducing the weaker statement from the stronger one), the derived function of

$$y + x^{[\ln \ln \sin^2(x + 1)]}$$

is not defined. We noted earlier in the chapter that a function is not necessarily defined by a formula, and now we know that a formula does not necessarily define a function (or, indeed, anything at all).

9

A Drop of the Hard Stuff: Analysis

Great fleas have little fleas upon their backs to bite 'em,
And little fleas have lesser fleas, and so *ad infinitum*.
And great fleas themselves, in turn, have greater fleas to go on;
While these again have greater still, and greater still, and so on.

Augustus De Morgan, A Budget of Paradoxes

The branch of mathematics called *mathematical analysis*, or usually just *analysis*, is the study of infinite processes, such as the limit of a function or the bite of the humble flea. Because some special properties of the real numbers are necessary for satisfactory results on limits, this means that analysis can be characterized as the study of the properties of the real numbers and the complex numbers (which include the real numbers), whereas algebra is more concerned with the rational numbers, and arithmetic (that is, number theory) with the integers. Elementary analysis, the part concerned with functions of a single real variable and with infinite series, has the best claim to be regarded as the most important and central part of mathematics, both for its own applications and for the support it gives to calculus. Unfortunately, the subject is not only important but also presents some initial difficulties, because the basic definitions are of a markedly unfamiliar kind. We shall take a brief look at these difficult definitions after we have considered some evidence for the importance of analysis.

In order to demonstrate that calculus still needs analysis even after its basic formulae have been proved, we start by studying the evaluation of the integral

$$I = \int_{-1}^{1} \frac{dx}{1 + x^2}.$$

By a calculus formula, the indefinite integral of

$$\frac{1}{1 + x^2}$$

is the principal value of the *inverse tangent*, $\tan^{-1}x$, that is, the value of θ such that $-\frac{1}{2}\pi < 0 < \frac{1}{2}\pi$ and $\tan \theta = x$. Therefore

$$\begin{aligned}
I &= \tan^{-1} 1 - \tan^{-1} (-1) \\
&= \tfrac{1}{4}\pi - (-\tfrac{1}{4}\pi) \\
&= \tfrac{1}{2}\pi.
\end{aligned}$$

Let us evaluate the integral again, with the aid of the substitution

$$x = \frac{1}{t}.$$

As $t = 1$ when $x = 1$ and $t = -1$ when $x = -1$, while

$$\frac{dx}{dt} = -\frac{1}{t^2},$$

by the Substitution Rule,

$$\begin{aligned}
I &= \int_{-1}^{1} \frac{dx}{1 + x^2} \\
&= \int_{-1}^{1} \frac{1}{1 + \dfrac{1}{t^2}} \frac{dx}{dt}\, dt \\
&= \int_{-1}^{1} \frac{t^2}{1 + t^2} \frac{(-1)}{t^2}\, dt
\end{aligned}$$

140

$$= -\int_{-1}^{1} \frac{dt}{1 + t^2}$$

$$= -I,$$

because the name of the variable in an integral does not affect the value. We deduce that

$$2I = 0$$

and, because $I = \frac{1}{2}\pi$, we conclude that

$$\pi = 0.$$

No matter what definition is used for π, this value is unconvincing. For example, if π is defined by the formula $A = \pi r^2$ for the area A of a circle of radius r, then the fact that a square of side $\sqrt{2}\, r$ can be drawn with its vertices on the circle means that $A > 2r^2$ and therefore $\pi > 2$. On the other hand, assuming that we have used the formulae of calculus correctly, the error must lie at a deeper level, in the foundations of calculus. In other words, the error that has been made belongs to analysis.

Because analysis is the foundation for calculus, it is tempting to suppose that errors in calculus like this could be entirely avoided by studying analysis first, but there are two good reasons for not doing so. The first is that the concepts needed in calculus are less general than those needed for analysis, so the experience of calculus is a useful preparation for analysis. For example, it is easy to grasp the meaning of 'sin $\frac{1}{4}\pi$' because it is merely a number, a value which can be obtained from a calculator or a book of mathematical tables. The idea of 'sin x' is only a little more difficult, as it is the value of the sine function for the arbitrary real number x. But what is the meaning of 'sin'? Indeed, does it exist at all? The first two of these ideas* are those used in calculus, though with some reference to the third, the sine function. However, the sine function would occur only as an example in an analysis course, which is more concerned with *classes* of functions, such as the class of all differentiable functions.

The second reason for starting with calculus is a consequence of the definition of a limit. Just as Eudoxus used inequalities to define the equality of real numbers in classical Greek mathematics, Karl Weierstrass and the other nineteenth-century mathematicians used inequalities to define limits. Unfortunately, the definition of the limit of a function is not a criterion concerning an inequality, but a sentence containing several inequalities. This is the source of the initial difficulties which are experienced in the study of analysis. These difficulties can be diminished by first gaining some familiarity with the ingredients of the defining sentences, that is, with inequalities, logic and functions. The prior study of calculus admirably serves the purpose of increasing familiarity with functions, as well as making its formulae available for applications as early as possible. It also mirrors mathematical history, because most of the results of calculus were known before a proper definition of a limit was found. However, we can now benefit from hindsight in constructing such a calculus course by including in it the statements of theorems of analysis when they are needed, but omitting their proofs.

After that, the initial difficulties in studying analysis can be reduced to a minimum by doing some work on logic and inequalities before tackling the analysis. Even this would not entirely remove the difficulties, because they are caused by unfamiliarity with the definition of a limit taken as a whole, but, for the same reason, the difficulties all disappear after a sufficient period of study. Afterwards, the memory retains a sense that there are difficulties present, but they can no longer be found.

Now let us work towards the definition of the limit of a function by analysing a particular example. As we showed in chapter 8, the differential coefficient of $f(x) = x^3$ at $x = 2$ is given by

$$f'(2) = \lim_{h \to 0} \frac{(2 + h)^3 - 2^3}{h}$$

and we can rewrite this as

$$f'(2) = \lim_{x \to 2} N(x),$$

142

where

$$N(x) = \frac{x^3 - 8}{x - 2}.$$

Because

$$x^3 - 8 \equiv (x - 2)(x^2 + 2x + 4),$$

the function

$$N(x) = x^2 + 2x + 4$$

provided that $x \neq 2$, but the formula given above for $N(x)$ reduces to the meaningless expression $0/0$ when x = 2, so the function $N(x)$ is defined by the formula

$$N(x) = \left\{ \begin{array}{ll} x^2 + 2x + 4 & \text{when } x \neq 2 \\ \text{undefined} & \text{when } x = 2. \end{array} \right.$$

The graph of $y = N(x)$ appears in figure 11, with a gap one point wide at $x = 2$ because $N(x)$ is not defined there. It is obvious that in order to avoid this gap, the definition of the limit of $N(x)$ as x tends to 2 must not use $x = 2$. On the other hand, $N(x)$ is defined for all other values of x, so why not take x to be approximately equal to 2? Before we try this, let us first explore the meaning of 'approximately equal' by reflecting on the following statements.

'These two stars are only a billion km apart, so they are almost in collision.'

'This proposal would only change the budget by a million pounds, which is chicken feed.'

'The architectural plans must be drawn to an accuracy of one millimetre.'

'The distance between these two particles is one millimetre, so they are very far apart.'

These statements serve to illustrate the following.

The Approximation Theorem Any two real numbers are approximately equal,

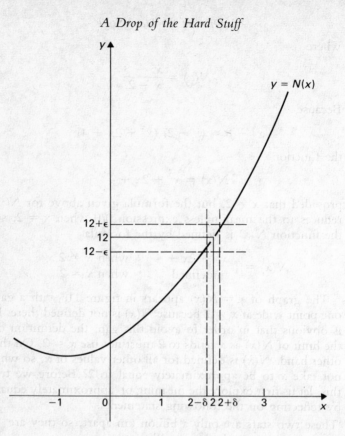

Figure 11 The limit of a function

under suitable conditions. This implies that the statement '*x* is approximately 2' is useless unless the accuracy of the approximation is also given or implied. We say that *x* is approximately equal to 2 with *accuracy k* (a positive real number) if *x* differs from 2 by less than *k*, or, in symbols,

$$2 - k < x < 2 + k,$$

which can be written more briefly as $|x - 2| < k$.

In a particular application of mathematics, it is often possible to determine an accuracy which is suitable for all the calculations. For example, in commerce, payments are

144

accurate only to the nearest penny. However, the idea of a limit is part of pure mathematics and so it must be applicable to all problems, both present and potential. Therefore the standard of accuracy must be good enough for all the physical problems envisaged by the hypothesis which Nicholas Odgers proposed in 1863. Odgers's Hypothesis asserts that every fundamental particle in our universe is a galaxy in miniature, whereas our galaxy (the Milky Way) is only a particle in some larger galaxy – Augustus De Morgan's 'flea poem' at the head of the chapter is his comment on this hypothesis. If we are to frame our definition of limit to cover Odgers's Hypothesis we must choose an accuracy that is satisfactory for use in all the chain of galaxies, so no real number would be an adequate choice. The only way we can be prepared for all the possibilities is to choose the accuracy to be a real variable, for which suitable values can be substituted in particular applications. Because the independent variable x can only be made approximately equal to 2, we can only expect to find values of $N(x)$ which are approximately equal to the limit L. We hope to find these values of $N(x)$ for values of x near 2 – indeed, we hope that $N(x)$ will be approximately L if x is sufficiently near 2. This leads us to the following.

Informal Definition $N(x)$ has the *limit L* as x tends to 2 if, for any accuracy that is required, $N(x)$ is approximately equal to L whenever x is sufficiently close to 2 but not equal to 2.

It is clear from figure 11 that if $N(x)$ has a limit L as x tends to 2, then $L = 12$. According to the informal definition, we need $N(x)$ to be approximately 12 with an accuracy which is a positive real variable, and this variable is traditionally denoted by ε. That is, we require

$$| N(x) - 12 | < \varepsilon$$

whenever x is sufficiently close to 2 but not 2 itself. If we measure how close to 2 the variable x is by means of a positive real variable δ (another traditional choice), then we can replace the statement that x is within δ of 2 but not equal

145

to 2 by

$$0 < |x - 2| < \delta.$$

This informal definition for this particular case can now be rewritten formally as follows.

Specialized Definition $N(x)$ has the *limit* 12 as x tends to 2 if, for each real number $\varepsilon > 0$, there exists a real number $\delta > 0$, which may vary with ε, such that

$$|N(x) - 12| < \varepsilon$$

for all real numbers x such that

$$0 < |x - 2| < \delta.$$

If we look at figure 11, we can see that this means that if we draw the pecked horizontal lines with $y = 12 + \varepsilon$ and $y = 12 - \varepsilon$, then we must be able to find a positive real number δ such that the graph does not pass out of the top or bottom of the box formed by these two lines and the pecked vertical lines with $x = 2 - \delta$ and $x = 2 + \delta$.

This specialized definition can be generalized into the following general definition if we add a preliminary sentence to define our terms.

Definition   Let a and $k > 0$ be real numbers and let $f(x)$ be a real function which is defined for all values of the real variable x such that $0 < |x - a| < k$. The *limit* of $f(x)$ is L as x tends to a, written

$$\lim_{x \to a} f(x) = L,$$

if and only if, for each real number $\varepsilon > 0$, there exists a real number $\delta > 0$, which may vary with ε, such that

$$|f(x - L| < \varepsilon$$

for all real numbers x such that

$$0 < |x - a| < \delta.$$

146

Now that we have obtained a definition for the limit of a function, we can, if we wish, prove result about limits like 'if the function $f(x)$ has the limit L and the function $g(x)$ has the limit M as x tends to a, then the product function $f(x)g(x)$ has the limit LM as x tends to a.' Once we are armed with these results, we can attack the problems concerning continuous functions which we mentioned in chapter 8.

Let $f(x)$ be a real function of the real variable x and let a be a real number. Then $f(x)$ is *continuous at a* if

1   $f(a)$ is defined,
2   $\lim\limits_{x \to a} f(x)$ is defined, and
3   $\lim\limits_{x \to a} f(x) = f(a)$.

There is not much that can be proved about a function which is continuous only at one point a, and the main results concern a function $f(x)$ that is *continuous in the closed interval* $[c,d]$; that is, $f(x)$ is continuous at a for all real numbers a such that $c \le a \le d$, where c and d are real numbers such that $c < d$. One example of such a result is that the function $f(c)$ has an unbroken graph in the closed interval $[c,d]$ if and only if $f(x)$ is continuous in $[c,d]$. Note that the unbroken graph of a continuous function need not be at all smooth – indeed, at many points there may be no tangent at all. To ensure that the graph of the function $f(x)$ has tangents at every point in the closed interval $[c,d]$, $f(x)$ must be *differentiable* in $[c,d]$, that is, the derived function $f'(x)$ must be defined for all x such that $c \le x \le d$. A function which is differentiable in the closed interval $[c,d]$ is always continuous in $[c,d]$, and therefore has an unbroken graph with a tangent at each point with x in $[c,d]$. However, such a function need not have a smooth graph because the tangents at neighbouring points on the graph may have unrelated gradients. In fact, a function $f(x)$ has a smooth graph in the closed interval $[c,d]$ if and only if the derived function $f'(x)$ is continuous in $[c,d]$.

Although its applications to calculus have made mathematical analysis an essential part of mathematical knowledge, the subject already existed before the discovery of calculus in the works of those whom Newton praised when he said 'If I have seen a little farther than others it is because I have stood on the shoulders of giants.' Among these giants was John Napier, whose major achievement has been celebrated by Diogenes O'Rell in the following words.

> John Napier, Baron Merchiston,
> Watched his logger fell some birches on
> His estate, and, seeing his skill and vigour with 'em,
> He discovered the logarithm.

As it is easier to add real numbers than to multiply them, an obvious objective in Napier's time was to find a method of multiplication which was performed by adding some related numbers. (In algebraic terms, the objective is to find an isomorphism between the group of positive real numbers under multiplication and the group of all real numbers under addition.) A clue leading to a method for achieving this objective is provided by the index law that asserts, for any positive real number b and any real numbers r and s, that

$$b^r \, b^s = b^{r+s}.$$

Multiplication can be executed by addition if we can find the *logarithms to base b*, $r = \log_b x$ and $s = \log_b y$, for the positive real numbers x and y such that $x = b^r$ and $y = b^s$. If so, by the index laws, we can calculate xy as $xy = b^{r+s}$.

From this it follows that

$$\log_b (xy) = \log_b x + \log_b y.$$

Consequently, to calculate $\log_b x$ is a consummation devoutly to be wished. But to do it: ay, there's the rub; for in that calculation what troubles may come? To evaluate b^r for varying r in order to construct a table of values leads to calculations that must give us pause. For example, to evaluate $b^{1.01}$ we must first calculate b^{101} and then extract its hundredth root. I leave the details of this calculation for any given b to the reader.

To find a method of calculating logarithms needed a stroke of genius, but it was probably preceded by a simple observation. Let us consider the sum to infinity of the geometric series with first term 1 and common ratio $-r$, that is, the series

$$1 + (-r) + (-r)^2 + (-r)^3 + \ldots + (-r)^n + \ldots$$

The formula for the sum to infinity gives us

$$\frac{1}{1 + r} = 1 - r + r^2 - r^3 + \ldots + (-1)^n r^n + \ldots$$

which can be interpreted as showing that the reciprocal of $(1 + r)$ can be found by calculating a sum, although it is the sum of an infinite number of terms. An examination of his papers might indicate otherwise, but perhaps Napier followed this clue by adding together various infinite sums rather like the geometric series. He then probably observed the following phenomenon.

Let x be a real variable and let

$$L(x) = x - \frac{x^2}{2} + \frac{x^3}{3} - \frac{x^4}{4} + \ldots + (-1)^{n-1}\frac{x^n}{n} + \ldots$$

Then

$$L(-x) = -x - \frac{x^2}{2} - \frac{x^3}{3} - \frac{x^4}{4} - \ldots - \frac{x^n}{n} - \ldots$$

If we add the two sums together term by term and write $n = 2m$ for the even-number terms, we find that

$$L(x) + L(-x) = 0 - \frac{x^2}{2} + 0 - 2\frac{x^4}{4} + \ldots + 0 -$$

$$2\frac{x^{2m}}{2m} + \ldots$$

$$= -x^2 - \frac{(x^2)^2}{2} - \ldots - \frac{(x^2)^m}{m} - \ldots$$

$$= L(-x^2).$$

Now, $L(x)$ cannot be the logarithm of x because $L(0) = 0$, wheres a logarithm must have $\log_b 1 = 0$, so we are led to consider the possibility that $L(x) = \log_b(1 + x)$. The equation we established for $L(x)$ then gives

$$\log_b(1 + x) - \log_b (1 - x) = \log_b (1 - x^2)$$
$$= \log_b \{(1 - x) (1 - x)\},$$

as required. Indeed, the *Naperian or natural logairthm* of x, denoted by $\ln x$, can be defined by the equation

$$\ln (1 + x) = x - \frac{x^2}{2} + \frac{x^3}{3} - \frac{x^4}{4} + \ldots +$$
$$(-1)^{n-1} \frac{x^n}{n} + \ldots$$

By evaluating $\ln x$ for a range of values, Napier would have been able to discover the base b of the natural logarithms by means of the relation $\ln b = 1$.

In fact, the base of the natural logarithms was more thoroughly examined by Euler, who showed that it was the *Euler number*, which is denoted by e and is given by the infinite sum

$$e = 1 + \frac{1}{1!} + \frac{1}{2!} + \frac{1}{3!} + \ldots + \frac{1}{n!} + \ldots$$

With the benefit of the work of Euler, Leibniz and Newton, we can define the natural logarithm more conveniently by

$$\ln x = \int_1^x \frac{dt}{t}$$

and we can derive the infinite sum for $\ln(1 + x)$ very easily as follows. By the substitution rule with $t = 1 + r$,

$$\ln (1 + x) = \int_1^{1+x} \frac{dt}{t}$$
$$= \int_0^x \frac{dr}{1 + r}$$

and, by the sum of the geometric series with first term 1 and common ratio $-r$,

$$\ln (1 + x) = \int_0^x (1 - r + r^2 - r^3 + \ldots$$
$$+ (-1)^n r^n + \ldots) \, dr.$$

If we suppose that, as for finite sums, the integral of an

infinite sum is the infinite sum of the integrals, then we obtain

$$\ln(1+x) = \int_0^x dr - \int_0^x r\,dr + \ldots + (-1)^n \int_0^x r^n\,dr + \ldots$$

$$= [r]_0^x - [\frac{r^2}{2}]_0^x + \ldots + (-1)^n [\frac{r^{n+1}}{n+1}]_0^x + \ldots$$

$$= x - \frac{x^2}{2} + \ldots + (-1)^n \frac{x^{n+1}}{n+1} + \ldots$$

By writing the term in x^n as the general term instead of the term in x^{n+1}, we reach the required formula

$$\ln(1+x) = x - \frac{x^2}{2} + \frac{x^3}{3} - \frac{x^4}{4} + \ldots + (-1)^{n-1}\frac{x^n}{n} + \ldots$$

This discussion has revealed how some logarithms can be calculated, but it also obliges us to ask a number of questions. First, how does one calculate an infinite sum? Indeed, what is *meant* by an infinite sum? In our calculations, we added infinite sums together term by term. Does that give a correct answer? We also integrated an infinite sum by integrating the terms. Is that valid? Would it also be correct to differentiate an infinite sum by differentiating the terms?

Any exposition of infinite sums needs to be accompanied by an exposition of a related topic, infinite sequences (sometimes called progressions), and it is more convenient to start with the latter. A (real, infinite) *sequence* is a set of real numbers arranged in a definite order, so that a term of the sequence corresponds to each positive integer. This statement is equivalent to the formal definition that a (real, infinite) sequence is a real function of a positive integer variable. In other words, for each positive integer n there is exactly one real number, the nth term of the sequence, corresponding to it. We proved at the beginning of chapter 5 that a sequence is not defined by a finite number of terms, no matter how large, so a general term must be defined. That is, if n is a positive integer variable (a symbol for which any positive integer may be substituted), the term corresponding to n, the *general term*, must be defined by a formula, by recursion, by a procedure for calculation or in some other definite way. We denote a sequence by

$$a_1, a_2, a_3, \ldots, a_n, \ldots$$

or, as the sequence is fully determined by the general term, we may refer to 'the sequence a_n', and the use of a subscript, n in this case, indicates that it is the positive integer variable. We are particularly interested in those sequences which have a limit. Roughly speaking, the sequence a_n has the limit L if, for any prescribed degree of accuracy, all the terms of a_n with n sufficiently large are approximately equal to L. This idea is formulated precisely in the following definition.

Definition Let $a_1, a_2, a_3, \ldots, a_n, \ldots$ be a real, infinite sequence and let L be a real number. Then the sequence a_n has the *limit L (as n tends to infinity)*, written

$$L = \lim_{n \to \infty} a_n,$$

if and only if, for every real number $\varepsilon > 0$, there exists a real number X, which may vary with ε, such that

$$|a_n - L| < \varepsilon$$

for all positive integers n such that $n > X$.

Because this definition is simpler than the corresponding definition for real functions, courses in analysis very frequently begin gently with a study of sequences and series before going on to study functions of a real variable.

An infinite sum is obtained by adding the terms of a sequence a_n to produce an *(infinite) series*: $a_1 + a_2 + a_3 + \ldots + a_n + \ldots$

(N.B. Sequences have commas, like $a_1, a_2, a_3, \ldots$; series have $+$ signs, like $a_1 + a_2 + a_3 + \ldots$.)

This can also be written in the contracted form

$$\sum_{n=1}^{\infty} a_n,$$

but both notations indicate, initially, only that a certain series is being investigated to determine whether it has a sum. For this purpose, we consider the *Nth partial sum* S_N of the series, where N is a positive integer variable, which is defined by

$$S_N = \sum_{n=1}^{N} a_n = a_1 + a_2 + a_3 + \ldots + a_n + \ldots + a_N.$$

This allows us to form the *sequence* of partial sums

$$S_1, S_2, S_3, \ldots, S_N, \ldots$$

and determine whether it has a limit. If the sequence of partial sums of the series

$$\sum_{n=1}^{\infty} a_n$$

has a limit, it is called the *sum* or the *sum to infinity* of the series and the sum is denoted by the same symbols as the series, that is, the sum is denoted by

$$\sum_{n=1}^{\infty} a_n = a_1 + a_2 + a_3 + \ldots + a_n + \ldots$$

Thus we have given a meaning to an infinite sum, but as with the limit of functions of a real variable and sequences, the sum of a series is not necessarily defined. If a series has a sum it is called *convergent*, otherwise it is *divergent*. A few series have sums which can be expressed algebraically, but most series that appear in applications are convergent series of which the sums are required in the applications but cannot be expressed in simpler ways. Our earlier use of a series to give the value of $\ln (1 + x)$ was an example of this. As a result of this use of series, the most important problem concerning a series is to determine whether it is convergent. For example, in electronic computation, very many functions are evaluated as the sums of series, so it is vital to know that the series are convergent in order to avoid interminable and erroneous calculations. However, the convergence problem has applications outside mathematics as well, such as the following one.

Problem

After Count Oktavian Rofrano married Sophie von Faninal (the Baron's former fiancée), Baron von Haerdupp deeply regretted his behaviour concerning the betrothal, which had been so far below his normal standard of rectitude and courtesy. He diagnosed his lapse as an emotional reaction to his bereavement at a time of financial anxiety, and he decided that his further financial efforts had better be confined to making suitable investments. For this reason he visited the Urbanian stock exchange, where he was offered an attractive gilt-edged investment bond. The initial premium for each bond was 100 spires, but it paid a dividend of the same amount after one year, so the bond virtually paid for itself. Furthermore, the dividend at the end of each year was so devised that at the end of the nth year the dividend was to be $100/n$ spires, while the renewal premium was only $1000/(n + 1)$ spires, leaving a net yield of $100/n(n + 1)$ spires every year throughout life. The modest but reliable (though decreasing) income accruing from these bonds persuaded the Baron to invest a large sum of money in them. How good was this investment?

Testing a series to decide if it is convergent somewhat resembles evaluating an integral in that there is a variety of tests available and the immediate problem is to choose the one most likely to be effective. However, in a large number of cases, the following simple test is the first to be applied.

The Divergence Test
If the sequence a_n does not have limit 0 then the series

$$\sum_{n=1}^{\infty} a_n$$

is divergent.

To apply the Divergence Test to a series, we calculate the limit of the sequence of terms. If this sequence does not have a limit or the limit is (say) 1, then we know that the series is divergent. On the other hand, if the limit of the sequence of terms is 0, we can reach no conclusion from the Divergence Test and we go on to apply some of the, literally, infinite number of other convergence tests.

In order to illustrate the calculation of the sum of a convergent series, let us consider the series for the natural logarithm:

$$\ln (1 + x) = \sum_{n=1}^{\infty} \frac{(-1)^{n-1}}{n} x^n.$$

It can be proved that this series is convergent if and only if $-1 < x \le 1$, therefore $\ln x$ can be evaluated for $0 < x \le 2$. We shall indicate how to use the series to evaluate $\ln x$ for any value of x and how to deduce from $\ln x$ the common logarithm $\log x$, that is, the logarithm of x to base 10. Let us start (for reasons that will be obvious later) by calculating $\ln$ 1·5. As this number is probably irrational, we probably cannot write down its exact value, but, in any case, the practical use of a series is to obtain a suitable approximation to its sum.

We shall calculate $\ln$ 1·5 correct to four decimal places, so we start our calculations using five decimal places and we write down all the terms of the series until they make no contribution to the fifth decimal place. By the Divergence Test, the terms of this convergent series must have limit 0, therefore, by the definition of the limit of a sequence, there is a term of the series after which all the terms are less than 10^{-5} so that (we hope) they will make no contribution to the fourth decimal place of the sum. Then

$$\ln 1·5 = \ln (1 + \tfrac{1}{2})$$

$$= \tfrac{1}{2} - \frac{1}{2} (\tfrac{1}{2})^2 + \frac{1}{3} (\tfrac{1}{2})^3 - \frac{1}{4} (\tfrac{1}{2})^4 + \ldots$$

$$+ \frac{(-1)^{n-1}}{n} (\tfrac{1}{2})^n + \ldots$$

$$= 0·5 - 0·125 + 0·04167 - 0·01563 + 0·00625$$
$$- 0·00260 + 0·00112 - 0·00049 + 0·00022$$
$$- 0·00010 + 0·00004 - 0·00002 + 0·00001$$
$$- 0·00000 + \ldots$$

The sequence of partial sums for this series is

0·5, 0·375, 0·41667, 0·40104, 0·40729, 0·40469, 0·40581,
0·40532, 0·40554, 0·40548, 0·40546, 0·40547, 0·40547, . . .

Accordingly, we deduce that

$$\ln 1·5 = 0·4055,$$

correct to four decimal places.

We can calculate $\ln (1 + x)$ at least as easily as this for any value of x in the range $-\frac{1}{2} \le x \le \frac{1}{2}$, but the series for $\frac{1}{2} < x \le 1$ require more terms to be calculated for a similar result. For example, the series

$$\ln 2 = 1 - \frac{1}{2} + \frac{1}{3} - \frac{1}{4} + \ldots + (-1)^n \frac{1}{n} + \ldots$$

has a sequence of partial sums in which the first digit in the decimal expression alternates between 6 and 7 from the fifth term of the sequence until the seventy-fifth term. It looks as if to calculate $\ln 2$ to (say) 10 decimal places by this method constitutes a lifetime's work.

If also illustrates that the simple numerical technique we used for $\ln 1·5$ will not necessarily work for an arbitrary convergent series. The study of techniques of this kind is the province of the important topic called *numerical analysis*, while lies inside the boundaries of both mathematics and computing. In our case, we can calculate $\ln 2$ much more easily as follows:

$$\ln 2 = \ln \left(\frac{4}{3} \times \frac{3}{2} \right)$$

$$= \ln \left(1 + \frac{1}{3} \right) + \ln \left(1 + \frac{1}{2} \right)$$

$$= 0·28768 + 0·40547$$

$$= 0·69315,$$

calculating $\ln (1 + 1/3)$ in the same way as $\ln (1 + 1/2)$. By adding $\ln 2 = 0·6932$ to $\ln x$ for $\frac{1}{2} \le x \le 1·5$, we can extend the evaluation of $\ln x$ to the range $\frac{1}{2} \le x \le 3$.

We could repeat this process for larger values of x, but we can use an alternative technique for finding $\ln (k + 1)$ in terms of $\ln k$ whenever $k > 1$. Because

$$\ln (k + 1) - \ln k = \ln \frac{k + 1}{k} = \ln \left(1 + \frac{1}{k}\right),$$

we can find $\ln (k + 1)$ by evaluating $\ln (1 + 1/k)$ by the series. For example, as $\ln 4 = 2 \ln 2$,

$$\ln 5 = \ln 4 + \ln \left(1 + \frac{1}{4}\right)$$

$$= 1{\cdot}3863 + 0{\cdot}2231$$

$$= 1{\cdot}6094,$$

when $\ln (1 + 1/4)$ is calculated by the series. Therefore

$$\ln 10 = \ln 2 + \ln 5 = 2{\cdot}3026.$$

To calculate $\log x$ when we know $\ln x$, we use the fact that

$$x = 10^{\log x},$$

from which we obtain, by taking natural logarithms,

$$\ln x = \ln (10^{\log x})$$

$$= (\log x) (\ln 10)$$

therefore

$$\log x = \frac{1}{\ln 10} \ln x.$$

In other words, to calculate the common logarithm $\log x$, we multiply the natural logarithm $\ln x$ by the number μ, where

$$\mu = \frac{1}{\ln 10} = 0{\cdot}4343,$$

correct to four decimal places.

Now that we have discussed the nature and calculation of sums of series, we can attempt to answer our other questions about series. The first one asks whether it is valid to add two series term by term, and it has an easy answer. In fact, its validity for two convergent series can easily be proved from the theorem that states that if s_n and t_n are sequences with limits L and M, then the sequence of which the general term is $s_n + t_n$ has limit $L + M$.

The questions concerning the validity of integrating or differentiating a series term by term can be answered together, but they are difficult to answer for series in general. However, these questions are more easily answered for the special kind of series that we have been considering. These are called *power series* and they are of the form

$$\sum_{n=0}^{\infty} c_n x^n = c_0 + c_1 x + c_2 x^2 + \ldots + c_n x^n + \ldots,$$

where x is a real variable which does not vary with n, and c_n does not vary with x. Every partial sum of a power series is a polynomial, so functions of a real variable which are sums of convergent power series can be regarded as the easiest generalizations of polynomials. The validity of term-by-term integration or differentiation of power series can be established for certain values of the real variable, but it requires a more advanced proof than the results we discussed earlier. This result justifies our use of term-by-term integration to obtain the series for $\ln(1 + x)$ from the geometric series for all values of the real variable x in the range $-1 < x < 1$.

Solution

Finally, we need to evaluate the profits that Baron von Haerdupp is likely to gain from his Urbanian investment. Let us suppose that he invested T thousands of spires in the gilt-edged bond. Then his net return in the nth year is $T/n(n + 1)$ thousands of spires, which is made up of a dividend of T/n thousands of spire and a premium of $T/(n + 1)$ thousands of spires. Therefore his total return after N years is S_N thousands of spires, where

$$S_N = \sum_{n=1}^{N} \frac{T}{n(n+1)}$$

$$= \sum_{n=1}^{N} T(\frac{1}{n} - \frac{1}{n+1})$$

$$= T[(1 - \tfrac{1}{2}) + (\tfrac{1}{2} - \tfrac{1}{3}) + \ldots + (\frac{1}{N} - \frac{1}{N+1})]$$

$$= T(1 - \frac{1}{N+1}).$$

Therefore, the profit in N years is $-T/(N+1)$ thousands of spires. Or, in other words, after N years the baron will still be $T/(N+1)$ thousands of spires out of pocket. However, because $T/(N+1)$ has limit 0 as $N \rightarrow \infty$, the series with Nth partial sun S_N has the sum T. This means that the baron will get his money back, but without interest, provided that he lives for ever.

The baron made a better investment when he bought a stout ladder for some roof repairs. It was promptly used by his elder stepdaughter when she eloped with a poor, local tenant farmer to Golders Green (they were not very good at reading maps). Although this did not increase the baron's income, it reduced his commitments and soon afterwards his prospects improved when his younger stepdaughter married a visiting English sportsman, who was the heir to some large estates. Unhappily, the optimism was short-lived, because his stepdaughter and son-in-law died while big game hunting and their baby son could not be found by the rescue party. As a result, all his son-in-law's inheritance passed to a more distant member of the Greystoke family.

10

A Compaignye of Sondry Folk: Mathematicians and their Lives

'I wrote not to be *fed* but to be *famous*.'

Laurence Sterne

Fame is notoriously fickle, and true fame must be based on great deeds with lasting effects. As William Hazlitt wrote:

Admiration, to be solid and lasting, must be founded on proofs from which we have no means of escaping; it is neither a slight nor a voluntary gift. A mathematician who solves a profound problem, a poet who creates an image of beauty in the mind that was not there before, imparts knowledge and power to others, in which his greatness and his fame consists, and on which it reposes. Jedediah Buxton will be forgotten; but Napier's bones will live.

Jedediah Buxton has, indeed, now been forgotten, but the logarithmic spirit of Napier's bones (a primitive slide rule) lives on vigorously in the memories of computers and pocket calculators. What kind of fame awaits those who seek it by orthodox mathematics? As a measure of the magnitude of mathematical fame, see how many winners of the Nobel Prize in Mathematics you can name in five minutes. This method actually exaggerates the height of mathematical fame because you belong to the minority with some interest in mathematics, or you would not be reading this book. And, you must admit, you failed to name a single prize winner.

But do not blame your memory. There *are* no prize winners, because there is no Nobel Prize in Mathematics!

The reason for this makes an interesting story. The leading Swedish mathematician at the beginning of this century was Gösta Mittag-Leffler (1846–1927), whose important work on complex analysis would have made him a leading candidate for a Nobel Prize, if one were instituted. The fact that Nobel did not institute such a prize has been interpreted as a wish to deprive Mittag-Leffler of such an honour. A persistent rumour suggests that there was a strong reason for such inaction, and this was identified as a love affair between Mittag-Leffler and Nobel's wife. This emphasizes just how clever Mittag-Leffler really was. Not only did he conduct this affair from Stockholm while Nobel lived in Paris, but he also overcame the great difficulty that Nobel was a bachelor.

However, the truth is far less strange than the fiction. Nobel instituted a prize for a subject only if he had a great personal interest in it, and he had no particular interest in mathematics. But mathematicians have not been excluded from Nobel's prize list, because Bertrand Russell (1950) and Aleksandr Solzhenitsyn (1970) both won Nobel Prizes in Literature. The prizes in mathematics which take the place of the Nobel prizes are the Fields Medals. About four of these are awarded at the International Congress of Mathematicians, which meets every four years. Unfortunately, the Fields Medal is not accompanied by a large cash prize, which is consistent with the low level of pay for mathemtical work. Indeed, it is claimed that the only mathematician who made a lot of money was Newton, and he only did so when he was Master of the Royal Mint. It follows that nobody should take up mathematics for wealth or fame, but only out of interest in mathematics which, paradoxically, might then lead to enduring fame among mathematicians. And this thought prompts the question 'What sort of people are mathematicians?'

WHAT ARE MATHEMATICIANS LIKE?

I have never been to Vladivostok, but, nevertheless, I feel sure that the mathematicians there form the same sundry mixture of people as anywhere else in the world, because they are similarly conditioned by the universal nature of mathematics. After a visit to Japan, Diogenes O'Rell expressed this thought as follows:

> Ev'rywhere on earth
> The mathematician finds
> universal truth.

However, all over the world, the mention of the word 'mathematician' brings to mind the eccentricities of some of the greatest. For example, when Archimedes' bath overflowed, he did not behave like any ordinary citizen by cussing and then asking his slave to mop up the mess. Instead, he leapt out of the bath and ran shouting down the street, which is why he is called the *Eureka Streaker*. Apparently he had a problem with an aqueous solution. Further, it is well known that Newton saw an apple fall in his orchard. He did not indulge in trivial reflections as to whether he had remembered to spray the tree or whether the apple was fit to eat. instead, he had thoughts of great gravity.

We are not here portraying only the greatest mathematicians, however, but we are trying to sketch a group portrait of the teachers, researchers and users of mathematics. What is this large collection of useful and hard-working people like? It has been said that the only rule is that

> Mathematicians are strictly irregular,

but there also persists a public image of a mathematician as a short, bald, bearded, bespectacled and absent-minded man.

The archetypal picture of a mathematician clearly corresponds very little with reality, and the reason is that it is a conscious creation, a deliberate physical representation of a human 'dessiccated calculating machine'. Actually, the last phrase was invented to describe Hugh Gaitskell, who was an economist and Chancellor of the Exchequer. Unlike him, mathematicians rarely rise to positions of power, though one of them, Eamon de Valera, became President of the Republic of Ireland. To discover an archetype of a mathematician which is less deliberately fabricated, it is necessary to find an image of a mathematician in the public's subconscious mind, and where can that be done more easily than at a party? The following is a typical party encounter between a mathematician and some person of local importance, whom we shall call 'The Dignitary'.

"Tell me", The Dignitary asks, with a smile blending condescension with distaste, which is not addressed to the large whisky he is holding, "What is your trade?"

"I am a mathematician."

"Oh . . ." gasps The Dignitary with his smug expression instantly replaced by a look of horror while he makes a barely successful attempt to prevent himself from stepping backwards, "I was never any good at that . . ."

This reaction reveals a different view of mathematicians. They are people of such superior intelligence that they are dangerous to know, and probably mad and bad as well. Perhaps this is why, in some science fiction novels, the evil genius who seeks to rule the world (or the universe) by the combined power of his mind and his machines is said to be a mathematician. This identification completely ignores the natural antipathy of mathematicians for machines, which is the reason why laboratory supervisors recognise only two ultimate tests for the durability of an apparatus: drop it over a cliff or let a mathematician use it. Perhaps the science fiction writer really meant that the evil genius was a computer scientist, who could certainly make and control machines with far-reaching powers. However, observations of computer scientists suggest that they are far too interested in improving the efficiency of their machines to be bothered with ruling the world. In fact, it is hard to forecast how an evil genius might start his career. After all, Adolf Hitler was an art student, while Francis of Assisi was a soldier. Only

163

one thing is certain: the evil genius would have an ardent desire to be supreme.

There is no doubt that some mathematicians are motivated by a similar competitive spirit, though they direct it with more moderation towards a better objective. But some others are far more notable for the collaborative skill with which they encourage others. A fine example of a mathematician like this was Issai Schur (1875–1941), who was Professor at the University of Berlin from 1919 until 1936, when he was dismissed by Hitler's henchmen. The long list of his research students and other collaborators includes many of the best workers in his subject (algebra) of this century, even though few of them worked on problems related to his own.

The organizers of the University of Cambridge examinations before 1924 took the opposite view of mathematics. They conducted the final year examination as a competition among those who were likely to gain first class honours (the *wranglers*), with the candidates listed in order of merit, starting with the Senior Wrangler. This arrangement of the examination was obviously bad for all the students who were not in the first class, but it provided stimulation for the best students. However, such a competitive regime has a less obvious disadvantage in that it depicts cooperation to solve problems as a form of cheating, whereas cooperation makes learning mathematics much easier, as well as being a good preparation for future industrial or research work. In addition to discouraging this useful 'cheating', a competitive system encourages a kind of institutionalized cheating. For example, students aiming at a high place in the list of wranglers would attend extra lessons with a *coach* or *crammer*, who would try to guess the questions for the coming examination and prepare the students for them.

Whether the majority of great mathematicians were of competitive or collaborative disposition is difficult to determine, particularly because the intrepretation of the evidence is likely to be influenced by the investigator's own views, in the same way that congenial proverbs can be chosen. As Diogenes O'Rell put it:

> The cream in the milk always goes
> To the top, as everyone knows,
> But I found when I looked
> At the jam as it cooked
> It was scum to the surface that rose.

If there is doubt about the attitudes to competition of other great mathematicians, there is certainly none concerning Pierre Fermat. Although he agreed in the letters to his correspondents on the importance of mathematical publication, he did not publish a single result, and he included only one proof in all his letters to other mathematicians. Otherwise, in a spirit of competition, he challenged his correspondents to find the proofs of the results that he claimed to have proved. As a result, all his work, except one theorem, would have been lost had not his son Clement-Samuel combed through his papers after his death and published all the complete results.

As a reward, Clement-Samuel Fermat is now probably found in that part of Heaven which is reserved for those who helped the work of the great mathematicians. Also there should be the wife of the hopelessly absent-minded Archimedes, as well as the sisters Klara and Elise Weierstrass who kept their similarly forgetful brother Karl in some semblance of order. Naturally, many of the parents of the great mathematicians are there, as well as Julie Dedekind (when she is not in the part of Heaven reserved for Jane Austen and the other novelists), who looked after her brother Richard throughout life, and Gauss's uncle Friederich Benz, whose astute questioning had provided such a good early training for his nephew. An area nearby is reserved for the teachers of the great mathematicians, and the Chair of Honour there is held by Isaac Barrow (1630–77) who, though a brilliant mathematician himself, resigned from the Lucasian Professorship at Cambridge so that it could be given to Newton.

It is less clear whether there is a place reserved in Hell for those who deliberately hinder mathematics and other branches of learning, because Dante did not investigate the matter when he toured hell, according to his poem *The Divine Comedy* (completed about 1320). However, it is significant that, although many mathematicians had lived by his time, he found none in Hell or Purgatory, so, perhaps, there are unsuspected advantages to be gained from studying mathematics. Maybe these advantages to be gained from studying mathematics. Maybe these advantages serve to counterbalance the incomprehension which mathematicians so often meet in their lives. For example, when The Dignitary recovers sufficiently from the shock of meeting a mathematician for the customary urbanity to return, the next step is sure to be the most obvious polite question to a mathematician:

'Are you pure?'

The bluntness of this question indicates the intensity of the public's subconscious doubts about mathematicians in certain intimate matters.

It is reported that only a tenth of American PhDs in Mathematics are women. It is puzzling that so small a proportion of mathematical researchers are women, even though women have long shown their ability in the subject and there is no obvious physical or social barrier against them. The lists of the mathematical specialists at one college were studied and it was found that women formed about 2/7 of the undergraduates, and 1/13 of the staff and postgraduate students. If these figures are at all typical, it would seem that factors outside mathematics cause many talented women to forego postgraduate work. Moreover, the percentage (about 30 per cent) of women mathematical undergraduates is well under the percentage (about 45 per cent) of girls in the schools, so there may be (external) reasons why more schoolgirls do not choose to study mathematics.

This imbalance is ever more striking because, historically, the mathematics departments were in the forefront of the movement to admit women to higher education. For example, Arthur Cayley led the attempt to admit women to the University of Cambridge in the nineteenth century. He had to be content with a compromise whereby they attended lectures and examinations unofficially. The mathematicians devised an equitable scheme in which the women wranglers (on the separate, unofficial, list) were listed as, for example, 'equal to the seventh wrangler', and the sincerity with which they operated this scheme was soon revealed when they listed one woman as 'above the Senior Wrangler'. This event provided inspiration for George Bernard Shaw's play *Mrs Warren's Profession*, in which the heroine is a mathematician. To take another example: in about 1915, David Hilbert tried to persuade the University of Göttingen to appoint Emmy Noether, probably the greatest of all women mathematicians, to a staff position from which she could be promoted to be a professor. Hilbert failed despite his plea that Emmy Noether's sex was unimportant because 'We are a university, not a bathing establishment.' Apparently, some of the other professors were worried by what they knew about Archimedes.

Before women were admitted to the universities, the only way they could gain a mathematical education was by taking private lessons, which was obviously a more expensive alternative. Ada Lovelace, the daughter of the poet Lord Byron, studied with Augustus De Morgan and later worked with Charles Babbage. Opinions are divided as to whether she contributed to Babbage's ideas for his proposed steam-driven computer, but she certainly published some work on analysis. Later in the nineteenth century there were university courses in mathematics open to the public in addition to degree courses, and Mary Ann Evans (1819–80, better known as the novelist George Eliot) attended such a course at the University of London's Bedford College. However, this was too late for Florence Nightingale, who was described as the best of all his private pupils by Joseph Sylvester (1814–97), who collaborated so notably with Arthur Cayley.

The time spent on mathematics by amateurs does not have the deep effect on the personality that a lifetime of mathematical immersion has on the professionals. Fortu-nately, these deeper effects include the development of some virtues. The best of these is patience. This is needed, at all levels of mathematics, first to study a problem in order to find a strategy for solving it, second to perform the detailed working to carry out the strategy, third to accept the disappointment when the strategy fails in its objective, and finally to return to the study of the problem in its new form. Of course, there is jubilation proportional to the effort expended if the strategy suceeds. Another virtue inculcated by mathematics is modesty, due to the limitation imposed by mathematical techniques. This helps mathematicians to avoid both the vaulting ambition and the consuming desparation depicted by Lewis Carroll:

> He thought he saw an Argument
> That proved he was the Pope:
> He looked again, and found it was
> A Bar of Mottled Soap.
> 'A fact so dread', he faintly said,
> 'Extinguishes all hope!'

In mathematics, results must always be proved, though in the easier cases the proof is in the form of a calculation. Consequently, the only kind of problem that can be considered in mathematics is the kind that can be written directly in a mathematical form, which immediately rules out questions like 'Which party will win the election?' or 'Do robins have spotted eggs?'. Further, even after several millennia of mathematics, it is not known how to solve some equations, others can be solved only in a limited number of cases and others can be proved to be insoluble. Consequently, only a minority of the problems which can be stated mathematically can actually be solved completely. Newton expressed this well when he summarized his life in the following words:

I do not know what I may appear to the world; but to myself I seem to have been only a boy playing on the seashore, and diverting myself in now and then finding a smoother pebble or a prettier shell than ordinary, whilst the great ocean of truth lay all undiscovered before me.

Leibniz shared Newton's discovery of calculus, but not his

views on mathematics. In 1666 he was looking for 'a general method in which all truths of the reason would be reduced to a kind of calculation' and some years later predicted that 'a few chosen men could achieve [this] within five years.' It must be admitted candidly that Leibniz was an optimist.

On the other hand, there are also some industrial diseases caused by mathematics, all affecting the mind. The first of them is that, after prolonged exposure to mathematical methods, the worker develops a tendency to overuse logical analysis.

In recent years, Edward de Bono has introduced the concept of *lateral thinking* as a means of combating the excessive use of logical analysis. Perhaps you would like to test your skill at lateral thinking with the following problem.

Problem

After Tarzan was discovered living in the jungle, he returned to England to claim the Greystoke estates, which he reorganized by sharing his spare manor houses among the members of the family, including his grandfather. Baron von Haerdupp was therefore able to give the Lerchenau estate to his surviving stepdaughter and move to England, where he enlivened his retirement by playing golf and breeding prize pigs. His pigs won the competition at the Barsetshire County Show for the best set of ten pigs, a competition for which there were some unusual regulations. These stipulated that the ten pigs were to be put into four equally sized pens which made up a square. The pigs were to be so arranged that, no matter how many times the judges walked clockwise round the pens, the number of pigs in each pen must be nearer to six than the number in the previous pen. There were no special regulations about the numbers of pigs in the pens, but the pens were only large enough for six pigs. How did Baron von Haerdupp arrange his ten pigs?

The second, and most serious, mathematical malady is the

development of a hard mind. *Hard-mindedness* is a subtle vice, which consists of maintaining that if a proposition is supported by a logically sound argument based on good evidence, then the proposition must be accepted. That is, a hard-minded person *insists* that what is true is actually true, but, of course, that is where the fault lies. The vice of the hard-minded person is not that of being right, but of not appreciating the psychological difficulties of others created by the fact that she is right. In order to reduce the efforts of their hard-mindedness, mathematicians should consider the possibility that there is some evidence against the proposition and also give their opponents in argument plenty of time to adjust their views in order to accept the proposition. However, despite all such efforts, Augustus De Morgan's words will probably remain true:

It is easier to square the circle than to get round a mathematician.

The mathematicians' third failing is to fall for the Fallacy of Categories, a tendency which they share with all other intellectuals. This fallacy, like so many others, is close to a sound principle, the *Principle of Elimination*, which was expounded as follows by that famous logician, Sherlock Holmes: 'When you have eliminated all which is impossible, then whatever remains, however improbable, must be the truth.' Holmes added, in an explanatory note; 'It may well be that several explanations remain, in which case one tries test after test until one or other of them has a convincing amount of support.'

Why do mathematicians sometimes misuse the Principle of Elimination? Perhaps this is because mathematics does not employ value judgements, so it is not an art. Further, mathematics does not employ the empirical method, so it is not a science. Consequently, mathematics does not exist and therefore any mathematician is a quasi-material form of self delusion. But, to prevent my disapperance by *reductio ad absurdum*, let me observe that my use of the Principle of Elimination was at fault because I did not ask 'Is the set of all intellectual disciplines other than the arts and sciences empty?'. The failure to prove that a classification covers all cases is the *Fallacy of Categories*. Of course, in our case, there are members of the set which I assumed was empty. These include all the medical sciences as well as the *logical sciences*, in which the method is that of logical deduction from axioms, a class which includes mathemtics, computer science, parts of philosophy and the theoretical parts of science and engineering.

A remarkably high proportion of mathematicians are appreciators of some form of music, and a large minority are performers as well, many achieving a very high standard of performance. It is widely held that this affinity of mathematicians for music is due to parallels of form in mathematics and music. Indeed, music of the baroque period (such as that by J. S. Bach) shares the jigsaw-puzzle neatness of the more elegant proofs in number theory and geometry, but it is hard to find any mathematical parallels in other music, which suggests that artistic similarity is not what attracts mathematicians to music. A more satisfactory alternative theory is that the qualities of mathematics and music balance in the mind. Mathematics is a form of meaning without intrinsic emotion whereas music is a form of emotion without intrinsic meaning, so they complement each other.

Some mathematicians have less ability as amateur musicians than as poets, and these include Joseph Sylvester, Sir William Rowan Hamilton and Augustus De Morgan. The astronomer and algebraist Omar Khayyam, who lived in Persia around 1100 AD, is now less remembered for his mathematics than for his poetry and, well, other things, just as he lamented:

> Indeed, the idols I have loved so long
> Have done my Credit in Men's Eye much wrong!
> Have drown'd my Honour in a shallow Cup,
> And sold my Reputation for a Song.

However, all the poetry about mathematics is written by mathematicians, such as Diogenes O'Rell, with the exception of some verses by Wordsworth. What is more, all mathematical humour is written by mathematicians, this

time without the asistance of Wordsworth, though some of
the humour of mathematicians has escaped into the outside
world, particularly in the work of Lewis Carroll and Alan
Milne. But there is one kind of joke that is a speciality of
mathematicians: the application of an idea to itself. Consider
the following example, which is by Diogenes O'Rell.

The Leader Speaks

The Anarchistic Party has
  Just triumph'd at the Polls,
And I, the Leader, can direct
  It to its chosen goals.

That crime should cease is our chief aim,
  So first we end its cause,
By legislation we will pass,
  Abolishing all Laws.

Administration only serves
  To hinder good intent,
And so we pledge to use our pow'rs
  To end all government.

And when we have impos'd on all
  These changes that we meant,
The People this whole land will rule
  By general consent.

But what do mathematicians do when they are not
working, listening to music or laughing? According to
general reputation, they spend much of the remaining time
watching or taking part in active sports, probably because
concentration of mathematics is intellectually very tiring and
therefore some enjoyable exercise (even if actually performed
by somebody else) provides welcome relaxation. For
example, Godfrey Hardy worked only four hours per day at
his number theory, then relaxed by playing real tennis, or
watching cricket or (in North America) baseball. Certainly
mathematicians need to choose between short concentrated
bursts of work or longer, less efficient, sessions; although
some major mathematicians seem capable of long, concentrated bursts of work. Considering how many mathema-

ticians take part in sport, very few have entered the top class. It is much easier to find mathematicians who are adept at intellectual games, and the most notable among these is Emmanuel Lasker, who was World Chess Champion from 1894 to 1921. The average mathematician is not quite in that class, but games like chess, bridge and backgammon take up many hours of mathematicians' time. This concludes our sketch of the nature of mathematicians, so now let us discuss where they are likely to be found.

WHERE DO MATHEMATICIANS LIVE?

Before we record observations like 'A mathematician's house has an integral number of storeys and walls meeting at right angles', let us first consider the kind of communities in which mathematicians live. We can start with the Palaeolithic (Old Stone Age) Folk who used to walk across North-sea-land to spend their summers at Clacton in about 200 000 BC. Their lives were spent following wild herds of deer and so forth, gathering berries, nuts, roots and herbs as they went. They had no domestic animals and, being always on the move, could store only as much as they could carry with them. Consequently, they had no use for any mathematics at all.

By about 2000 BC, the sea had closed round the British isles, but this did not prevent the arrival of the Beaker Folk, who appear to have been somewhat mobile, presumably to find pasture for their sheep and cattle and to find minable veins of the copper and gold with which they worked. However, they also built round hut near the fields in which they grew barley. As they could make beer from the barley and take milk from their cows, it is not known what they drank from their distinctive pottery beakers. But we do know how they made a date: they built a henge (such as Stonehenge). The daily life of the Beaker Folk required an ability to count sheep and the erection of henges required not only some ingenuity and skill at working stone, but also an ability to measure and communicate lengths. They could measure the lengths by counting paces, but they would require names for the numbers in order to be able to tell others about them. Such a community would have no need for a mathematician, but it had already taken the firt step along the mathematical road.

Many more paces had been taken along that road by the time Julius Caesar first landed in Britain in 55 BC, when it was occupied by various Iron Age people. Let us consider a farming village of that time in order to tell an everyday story of country folk. In their fields they grew wheat and barley (which they stored in pits), they kept hens, sheep and cattle around the village (at least, most of the time), they wove and dyed cloth, they worked wood with lathes and other tools, they built fairly large circular huts and they baked bread. Most of these activities require simple arithmetic in the form of counting and addition, and it is possible that the Iron Age Folk did some surveying, for which they would need multiplication as well. This kind of community would need many members to be capable of the simpler arithmetical operations and, perhaps, one or two members who could do some surveying. However, there was still neither the need nor the opportunity for anybody to do any more advanced mathematics, especially as the community had no method of writing, so all records had to be in symbols or in the memory.

Hence there would be no mathematicians to greet the Romans when they arrived, but the latter soon brought some of their own to direct surveying operations and to act as engineers and architects. Except for a few dark periods of history, mathematicians continued to be employed on large scale engineering projects until the seventeenth century (and beyond), and then there were founded the first posts specifically in mathematics, as professors at the universities. Since that time, society's need has grown rapidly for mathematical specialists, for teachers of mathematics and for a higher level of mathematical ability in the population at large. We can therefore conclude that any community which does large-scale engineering will have mathematicians, the larger communities of this kind will contain some mathematical specialists and a highly industrialized community needs mathematical specialists in large numbers.

At a given place in an industrialized community a certain minimum number of mathematicians is needed for good progress. For example, Sir William Rowan Hamilton had few mathematical companions in Dublin and consequently received limited mathematical stimulation from them. Unfortunately, this caused him to waste his time on a number of unprofitable ventures instead of the major problems he alone at that time was capable of solving. Their need for company forces mathematicians to work in cities, which conflicts with their need for peaceful surroundings conducive to contemplation. Consequently, mathematicians tend to live in the leafy suburbs where they can work without interruption. For example, from my study window I can hear the rustling of the leaves in the wood beside the house, the bubbling song of the curlews in the fields beyond the wood, the rushing of the river over the weir in the valley below and the roaring of the jumbos as they take off from the airport. This represents the natural place for all mathematicians, suspended between the ideal and the real. However, although we know where to find existing mathematicians, how are we to find the increasing numbers of mathematicians needed for the future?

HOW TO FIND FUTURE MATHEMATICIANS

Intelligence is an important component of mathematical ability, but it is not all of it, as some very intelligent peole are quite unable to cope with even elementary mathematics. There is often a simple reason for this: at some stage of their mathematical education they got left behind and, as nobody went back for them, the rest of the course might just as well have been delivered in Martian. That such people miss opportunities due to their lack of mathematics is sad, but there is an obvious remedy: go back to the point where they got lost and try again, more slowly. A phenomenon which is more significant for our purpose is the occurrence of people who find elementary mathematics easy, but then come to a point which they cannot pass, despite adequate help. The cause of this is not some external factor but, presumably, a shortage of an intellectual or moral quality needed for further progress in mathematics. Let us look at a few of these qualities without attempting to assess their relative importance.

It becomes vital, after the elementary stages of mathematics, to be an intellectual explorer, prepared to try new paths through the difficulties posed in the problems. As a mathematics course progresses, the methods taught are first used singly to solve the problems, then used in sequences. Later they are combined to produce new methods, and finally, in mathemtical research, completely new methods may be required, and even the problem is written in a definite form only after some exploration.

Another valuable quality is an ability to recognize patterns, which can be used by a mathematical explorer as hints for solving problems. Examples of the kinds of pattern which occur in mathematics are symmetries in geometrical figures or matrices or between variables in a formula, appearances of known formulae in larger formulae (such as the discriminant in the solution of the quadratic equation), rhythmic appearances of terms in a sequence, and formulae which are unchanged by a change in sign of the variable.

The ability to appreciate a pattern merges with the mathematically valuable ability to *generalize*, that is, to pass from a particular example to a general formulation, which can then be examined further. A particular form of generalization is *abstraction*, such as moving from an arithmetical formula to an algebraic formula. It is very important for a mathematician to have some ability to obtain more abstact results in this way, as well as to grasp the particular meanings of a result presented in the abstract form. In order to maintain interest in the more abstract forms of mathematics, it is necessary to be able to appreciate their more concrete consequences.

In addition to these intellectual properties, a mathematician also needs some moral qualities. We have already discussed the need for patience, a virtue which is rarely a gift. However, patience can be developed by persistence, which is certainly a more intrinsic property because even some babies show it. Also, some mathematicians have an intense desire to *know* the truth, rather than to be told it, and so they feel the need to *prove* as much as possible. However, it does not seem likely that this quality is necessary for engaging in mathematical work.

Do mathematicians have a distinctive mode of thought? If so, potential mathematicians could be detected by psychological tests which determined whether they had or could develop these characteristic thought processes. Unhappily, although mathematicians have been the subject of more psychological studies than any other group except writers and madmen, very little is known for certain about their method of thought. Various interviews with great mathematicians have elicited that, when searching deep in the mid for ideas, they encounter vague images resembling pictures or movements, which they then try to express in symbols or words. This had led to an opinion that 'great mathematicians think in pictures', but my own limited experience suggests that this opinion misinterprets fumbling words which express thoughts on the edge of the subconscious.

To test this idea, I have conducted a slight experiment into this problem. The method consists of walking up to a chosen person and saying 'Cow!'. It is surprising that the name of so peaceful a creature should produce such violent initial reactions, but they do not prevent the experiment from continuing with questions like 'What colour?', 'Where was it?', etc. A number of non-mathematicians have given answers like 'It was a brown cow, grazing on the green, green grass in a field with the five-barred gate open on a sunny day with . . .'. In other words, their reaction to the deliberately visual clue was to see a *picture*. Perhaps I should have had one of them with me when I tried to describe a cow to Zxt. Many mathematicians, as well as other people, replied 'The cow had no colour, it was just of colourless background.' In other words, they saw a *diagram*, which represents an element of abstraction in their thought. There were various kinds of exception: some saw the word 'cow' (so they probably think largely in words), some thought of something associated with a cow (such as milk), the yachtsman saw a lot of Cowes and the milkmaid saw an udder view. Perhaps the experiment should be continued with an ambiguous word like 'plane'. Also, one could investigate the suggestion that mathematicians who think in diagrams work by staring impassively into space while those who think in words scribble feverishly on scrap paper.

These two modes of thought are not the only ones relevant to mathematics. Because mathematics is concerned with pattern, there is considerable scope for using the regularities of motion, that is, kinaesthetic thought. Some understanding of kinaesthesia can be obtained by watching a drummer bet out a complicated rhythm. She keeps her drum beats in time by hitting the air on some of the silent beats, because this enables her to maintain a steady bodily rhythm.

Of course, we have not discussed all possible modes of thought. For example, none of the modes we have considered would enable one to remember a scent, yet such a memory is so powerful that there is a special word (*redolent*) relating to it. Its power also motivated the Duchess in *Alice in Wonderland* to say 'Take care of the scents and the sounds will take care of themselves.' (Here I have yielded to the mathematician's frequent temptation to quote Lewis Carroll. We should certainly resist this temptation and write in a simpler style by taking the advice: 'Begin at the beginning, and go on till you come to the end: then stop.')

Our ignorance of the thought processes of mathematicians contributes to the reliance of most elementary examinations of mathematical ability on tests in speed of execution. These tests do pick out some kinds of mathematician, but they fail to identity potential mathematicians who are rapidly bored by repetitous work designed to increase speed or who think slowly and deeply. And yet, tests of this kind are frequently used in senior schools to select an 'express stream', which takes mathematics examinations a year or so early. This would matter little if nothing further depended on this selection, but the same schools often only permit the members of the express stream to take the extended mathematics syllabus (intended for potential university students of mathematics and related subjects) in the final years at school. Worse still, some schools only allow members of the express stream to take mathematics at all in the final years. If this trend continues, the only students permitted to take mathematics in senior schools will be those who went to the special mathematical kindergarten! Early selection like this ignores the large number of students who change their subjects or their intended careers. This phenomenon was illustrated at a dinner party attended by eight academics who professed a wide range of subjects. It was discovered that everyone present had changed subject after leaving school. The earliest change was between school and university, the latest was after doing some postgraduate work in the former subject. (I shall not explain how I found *my* way from geography to mathematics.)

A factor which leads to the early selection of potential mathematicians is the existence of true mathematical prodigies. For example, Evariste Galois and Niels Abel made outstanding discoveries even though Galois died after fighting a duel at the age of 20 and Abel died of tuberculosis at the age of 27. Had Newton and Weierstrass, respectively, died at these ages, they would have discovered nothing, because Newton did not start his mathematical studies until he went to university at the age of 19 and Weierstrass studied law at university, though he failed and then started studying mathematics when he commenced his training as a teacher at the age of 24. Weierstrass than made a slow start in the subject, but Newton progressed very rapidly and by the age of 26 had made the discoveries that kept most of the mathematicians in the world busy until Gauss came along with further ideas.

Gauss, however, was a prodigy – indeed, he was two prodigies. Like Hamilton and Pascal, he was a true mathematical prodigy, who could do genuine mathematics so well in advance of his age that he strained the mathematical resources of those who were trying to teach him. True mathematical prodigies always succeed in life, but not necessarily in mathematics. For example, John Wilson (1741–93) is remembered for the theorem he proved while he was a student, but his success in life (for which he was knighted) was as a judge. The qualities (such as high intelligence) of a true mathematical prodigy seem to be readily transferable to other activities, whereas those of a musical prodigy are not. A second kind of mathematical prodigy is called an algorist. An *algorist* can manipulate mathematical formulae subconsciously and produce results which he cannot explain. Of course, one has to know something about mathematics in order to use this gift, but with its aid one rapidly runs ahead of proven results. If not a strictly mathematical prodigy, an algorist is someone with a rare and unquestionably valuable gift. Carl Jacobi (1805–51) was notable as both a mathematician and an algorist, and a good modern example was Srinivasa Ramanujan (1887–1920), who had a very productive collaboration with Godfrey Hardy.

There are least two further kinds of supposedly mathematical prodigy. One kind has visual intuition, so can grasp the spatial relationships between the parts of a complex geometrical figure. Unhappily, such a person can rarely explain these relations to anyone without comparable intuition, so the gift is not as valuable as it sounds. However, someone who has visual intuition and also true mathematical or manual or organizational ability should become a good geometer or artist or architect. The final kind of prodigy that we shall discuss is the arithmetical prodigy. Gauss was an arithmetical prodigy – indeed, he joked that he could calculate before he could talk. In his day, this gift enabled him to do mathematical work which others could not manage due to time problems, but this kind of problem could be done by any mathematician nowadays, provided he had a suitable computer. The fact that the arithmetical gift functions almost subconsicously, making great use of the memory, is the reason why arithmetical prodigies are not necessarily mathematically gifted. In fact, some arithmetical prodigies have been idiots (monomaniacs). Others have been otherwise normal people, and some of these have used their arithmetical gifts as part of effective variety acts. However, no matter what gifts a potential mathematician may have, he still needs to cultivate them, so let us now discuss how mathematics should be studied.

HOW TO STUDY MATHEMATICS

The study of mathematics, like that of any other subject, is governed by the following fundamental law.

The Inverse Military Law General Study is useful but Major Study is more important and Private Study is the most important of all.

If we accept this background advice, we then must ask 'how should one study *mathematics*?', and we come to the first, and most import piece of advice:

Don't

without an *interest* in mathematics. This interest might be because the mathematics is needed for some other study (such as engineering), but this should suffice as motivation provided that the connections between the mathematics and the other study are kept in mind. On the other hand, if the subject to be studied is mathematics itself, then only a pure interest in at least some of the mathematical topics is necessary. The 'wrong' topics can be endured, and possibly even enjoyed, for their relations with the interesting topics, or simply as the stone that necessarily comes with the peach. Either way, it is essential to fill one's life with what one enjoys. No, no. Not sleeping. To obtain the food to make sleeping comfortable, one has to work, but, happily, there is no rule which says that work must not be enjoyable.

However, there is a rule, imposed by the subject itself, which determines how mathematics should be studied: *steadily*. This is because studying mathematics is largely a combination of comprehending abstract ideas and assimilating techniques, both of which demand time for the mind to operate on them properly. On the other hand, working steadily can be done in a variety of ways. Some students work on each item as it occurs, others organize the week so that a particular topic is studied on (say) a particular day and some others prefer to concentrate on one topic at a time by giving each one (say) a week of its own. Whichever arrangement is preferred, the basic method is to work at examples, which should include many routine calculations as well as (or as a preliminary to) solving problems. In addition, an attempt should be made to understand each proof by finding its main step – of course, a long proof which goes through several stages will have several important steps. And, of course, the more all this work is discussed with others, the better, because what is fun for one can be more fun for two and, anyway, anyone who needs no help is in a good position to give it to others.

No matter how well one organizes one's work, however, sooner or later one will meet an item which repels all efforts to understand it. The usual reason for the difficulty is that one has not fully understood some earlier item, so the first move should be to reread the earlier work which relates to the difficult item. If one's own efforts with textbooks do not reveal the cause of the problem, consult a teacher of mathematics. If the problem still persists, mental staleness should be diagnosed. The treatment prescribed for this ailment is to rest the mind by using at least a few days for physical activity, such as practising one's backhand or exploring the countryside. It is at the end of this break that one's interest in mathematics is important to provide a call back to mathematical work.

Solving problems is an important part of mathematical education, but it is even more important in mathematical research and the applications of mathematics. To some extent, solving problems is a craft, to be learned by serving an apprenticeship, but the techniques used for finding solutions have been studied in their own right, and the results recorded in books on problem solving. To read such a book is no substitute for the apprenticeship, but it can certainly provide some useful guidance.

Unfortunately, no matter how efficiently one organizes

one's work and how well one solves problems, there is one evil which afflicts every student on a mathematics course. This evil is usually called 'the examination'. It is a gross error to assume that the course is a preparation for the examination – really, the examination is only there to test whether the course has been satisfactorily completed. All the same, the examination can haunt like a ghost whose threatening presence drives all the pleasure away from the course, so why should it not be exorcised? Alas, if the exorcist laid all such ghosts to rest, the country would eventually have no mathematicians of attested quality. This vision is not as immediately terrifying as a country without attested doctors and engineer, but it could readily become the scene for some spectacular disasters. So, instead of calling for the exorcist, it is better to prepare a welcome for the ghost and thereby transform it into a friendly one.

The first step towards such an amicable meeting is to work steadily through the course, but the pleasure of that final tryst could still be spoilt by ineffective revision and the wrong examination technique. This applies particularly to mathematics examinations because they require *different* techniques for revision and examination working from other subjects. The early part of the revision for a mathematics examination is little different from that for other subjects: just work (or rework) some examples on each item as well as draw up lists of methods, formulae and important theorems (with the main points of their proofs). Naturally, the questions in (say) the last paper for the examination should be worked through, as they should for any subject. It is the later stage of the revision that is so markedly different. About a week before the examination change to spending only a short time each day reading through the revision list, and spend the rest of the time preparing the body and mind for the rigours of the examination room by enjoyable exercise, relaxing mental pursuits and refreshing sleep. However, at this stage one should certainly not work any examples (hurried attempts usually fail, producing a needless lack of confidence) and on no account should one test one's memory for formulae, etc. Such a memory test is bound to fail, because one recalls mathematics bit by bit as one works, not as a complete entity, so the consequence of such a test is a sense of panic which even the most terrifying spook would be pleased to induce.

Now let us suppose the candidate has arrived in the examination room both fit and mentally fresh, ready to tackle problems and 'bookwork' proofs, which are really only problems of which the solutions have already been studied. (As a contrast, a history examination consists of factual questions or the writing of reasoned essays, whereas some other subjects, such as physics, have examinations which mix the styles for history and mathematics.) In mathematics examinations, there is no need to attempt the maximum number of questions, because marks accumulate as the work progresses: so much work, so many marks. A rough classification of mathematics examination papers distinguishes three kinds: papers with many short (perhaps 'multiple choice') questions, paper with longer, more problematic questions and papers with both short and long questions.

The technique for the short-question paper is to work through the questions in the order in which they appear, ignoring every question for which the method is not immediately obvious. When the end of the paper is reached, go back to the beginning and study the previously omitted questions until time is called. There is no need to consult the clock at any time during the examination. In the long-question paper, the required strategy is to attempt, at any moment, the unattempted question that one finds *easiest*. If a question turns out to have hidden difficulties, abandon it, and attempt the easiest remaining question. Continue in this way until time is called. Again, there is no need to consult the clock. The mixed-question paper, though, requires more judgement. Let us suppose that the paper consists of Section A containing many short questions and Section B containing the problems. Usually, the paper gives some advice like 'Spend about one third of the time on Section A.' However, personal differences are important in this matter, and some candidates will want to apportion their time differently. But part of the technique for such a paper is fixed: start with Section A, giving it the prejudged proportion of time but otherwise treating it like a short-question paper, then spend the rest of the time on Section B, treating it like a long-question paper until time is called. And after a few short but eventful years, the invigilator will call time in the final examination, and the student will have become a mathematician, ready to take up her share of the world's work.

MATHEMATICIANS AT WORK

Not so many years ago, there was a popular belief that 'Mathematicians only teach people to become mathematicians who only teach people to become mathematicians who . . .'. The basis for this strange view was that the senior school mathematics syllabus was dominated by Euclidean geometry and gave no hint of any further meathematical topics beyond it. This left a strong impression that mathematics was fully worked out and totally useless, except for the elementary algebra and arithmetic which everybody knew. Of course, those who only studied the more elementary parts gained the impression that mathematics consisted of 'harder sums' about a wider range of subjects. And, given a considerbly extended meaning of 'sums', that is not a bad image of the work done by mathematicians in modern scientific and technological research, with the mathematicians working as members of teams with scientists, computer scientists and engineers.

At a somewhat lower level of difficulty, many mathematicians work on organization problems of commerce and industry, making important decisions about the daily work. These mathematicians use techniques which rarely find their way into degree courses because they are not powerful enough to interest applied mathematicians or coherent enough to form a theory for the pure mathematicians. Perhaps they need to be put together to form their own branch of mathematics, though it would never prosper if it were given an honest title such as *grotty mathematics*. It would be much better to give it an elegant foreign name such as *Langshäftermathematik*.

However, despite all the posts for mathematicians in industry and commerce, there are mathematicians (though not enough of them) who choose to teach in schools and colleges. Teaching is not the best paid work, but it has other consolations, as Robert Bolt argued in his play *A Man for all Seasons*:

Sir Thomas More: . . . Why not be a teacher? You'ld be a fine teacher. Perhaps, a great one.
Richard Rich: And if I was who would know it?
Sir Thomas More: You, your pupils, your friends, God. Not a bad public, that.

For anyone with a good grasp of the subject, mathematics is a relatively easy subject to easy subject to teach because it is easy to recognize precisely what needs to be taught. However, it is essential to give the highest priority to teaching only what is true. In any subject, but especially in mathematics, once an untrue statement has been learned it can only be corrected by reading a suitable antibook. An *antibook* is like an ordinary book except that it contains *antiknowledge*, which induces mutual destruction whenever it meets the relevant knowledge, just as physical particles and antiparticles destroy each other. Unfortunately, no effective antibook has yet been written – perhaps it will be unwritten some time recently. This is why the truth must have precedence over the clarity and generality of what is taught. For example, it is adequate to start learning calculus for functions with smooth graphs and later to make this idea clearer by giving a precise criterion for such functions. Later still, more general theorems can be proved in mathematical analysis which include the original calculus results. It is easy to clarify the detail of a known correct result and then add further results to it later, but it is very difficult to appreciate that an incorrect principle that one has believed for a long time is actually wrong. Of course, the teacher also needs to bear in mind the following.

The Teacher's Golden Rule The teacher should inspire sometimes, enlighten often, and encourage always

The modern idea which has replaced the belief that mathematicians only teach is that all non-teaching mathematicians work with computers. This idea has some merit, in that many mathematicians are employed on work with computers. Such work makes use of mathematics, whereas, despite the world shortage of mathematicians, many of them work in professions that make no direct use of mathematics, such as accountancy and law, where the clear thought that is necessary for mathematics is valuable. Clear thought is also useful for solving the kind of problem that the Barsetshire County Show posed for pig breeders.

Solution

Perhaps inspired by reading the works of Lewis Carroll, Baron von Haerdupp started modestly by putting 1 pig in the first pen he came to, and, moving clockwise round the pens, he put 3 pigs in the second pen because, clearly, 3 is nearer to 6 than 1 is. Into the next pen he put 6 pigs, which is satisfactory because 6 is nearer to 6 than 3 is. As he had then used up his supply of pigs, the baron left the fourth pen empty and thus completed the arrangement because, undoubtedly, nothing is nearer to 6 than 6 is and, passing to the first pen of the circuit, he noted that 1 is nearer to 6 than nothing is. With this arrangement, the baron's pigs won first prize.

However, rather disappointingly, among all the prizes that his pigs won, there was never a first prize for his fat sows, because his neighbour Lord Emsworth always won this class with his famous sow, the Empress of Blandings. Nevertheless, the baron's pigs were strong and healthy, and they became the ancestors of the splendid pigs that Dan Archer reared at Brookfield Farm in Ambridge. But, of course, that was after the baron's golden evening of par-scores and pig-swill had ended and he had been laid to rest in the remote country churchyard, where his grave can still be found in the corner between the church and the motorway. There, the rain beats down on his gravestone, which is engraved with the following epitaph by Diogenes O'Rell.

Von Haerdupp in fame does outshine
All the barons of his noble line,
Not for golf on the links
Nor for other high-jinks,
But for breeding all Dan Archer's swine.

Further Reading and References

The books listed under each heading below are mainly intended to be borrowed from libraries to provide more information and explanations than are possible in a short book like this. Some of the books which are indicated as 'introductory' might be worth buying to help with independent study, but the others should be borrowed and dipped into to find out what the topic concerned is like. This remark particularly applies to the books described as 'textbooks', because such books go too far and too fast for unaided study and, in any course on the topic, a different textbook will probably be needed.

At the end of the entry for each chapter, details of where the quotations in the chapter can be found are given, provided that some information other than that in the text is available.

GENERAL

A book with objectives very close to those of this book is *Prelude to Mathematics* by W. W. Sawyer (Pelican series paperback; Penguin, Harmondsworth, 1955), which is a sound but somewhat old-fashioned introduction to mathematics. A good elementary introduction to mathematical ideas is W. W. Sawyer's *Mathematician's Delight* (Pelican series paperback; Penguin, Harmondsworth, 1943) and a much more advanced introduction is *Concepts of Modern Mathematics* by Ian Stewart (Pelican series paperback; Penguin, Harmondsworth, 1975). A useful introductory textbook, which would be very suitable for anybody who wished to brush up their mathematics before starting a course, is *An Invitation to Mathematics* by Norman Gowar (Oxford University Press, Oxford, 1979).

CHAPTER 1

An alternative view of mathematics is put forward by Philip J. Davis and Reuben Hersh in the first half of *The Mathematical Experience* (Pelican series paperback; Penguin, Harmondsworth, 1984). A non-technical, and therefore somewhat simplified, account of contemporary mathematics is given in the set of essays *Bridges to Infinity* by Michael Guillen (Rider, London, 1984). These essays are commendably indifferent to details of mathematical history and careless concerning the details of some definitions and theorems, but they contain very clear discussions of the appliction of mathematics to physical problems, especially in the essays 'Nothing like common sense' and 'An article of faith'.

The need for higher mathematics in engineering is demonstrated by Eric Laithwaite in pages 62–8 of *Invitation to Engineering* (Basil Blackwell, Oxford, 1984).

The value of mathematics in economics is discussed by David Whynes in pages 28–32 of *Invitation to Economics* (Basil Blackwell, Oxford, 1983).

The quotation from Hilaire Belloc forms the third paragraph of his essay 'On Cheeses'.

The quotations from the works of William Shakespeare are *Hamlet*, Act III, scene i, line 56 (written in mathematical symbols) and *Macbeth* (I, iii, 140–2), quoted by Zxt.

The first item of anonymous verse has been ascribed tentatively to Diogenes O'Rell (who, of course, is familiarly known as 'Dog'). There does not appear to be a manuscript or any documentary evidence to connect him with it, but its style indicates that it is an example of Dog O'Rell verse.

CHAPTER 2

Axiom systems are discussed by Ian Stewart in pages 113–20 and 76–8 of *Concepts of Modern Mathematics* (Pelican series paperback; Penguin, Harmondsworth, 1975). The axioms of Euclidean geometry are discussed in greater detail by E. R. Stabler in pages

26–32 and 232–8 of *An Introduction to Mathematical Thought* (Addison-Wesley, Reading, Massachusetts, 1948). An introductory textbook on logic and axiom systems for algebra is *The Foundations of Mathematics* by Ian Stewart and David Tall (Oxford University Press, Oxford, 1977).

The epigraph is a (variously reported) remark by Archimedes about the force that can be obtained by using a lever.

The quotation from William Wordsworth's *The Prelude* has been selected from lines 75–114 of Book V.

CHAPTER 3

E. R. Stabler discusses scientific method and the axioms of mechanics in pages 102–19 of *An Introduction to Mathematical Thought* (Addison-Wesley, Reading, Massachusetts, 1948). *Elementary Mechanics* by D. A. Quadling and A. R. D. Ramsey (G. Bell, London, 1972) is a good elementary textbook. *Principles of Mechanics* by John L. Synge and Byron A. Griffith (McGraw-Hill, London, 1959) is a more advanced textbook which has an interesting section on the founations of mechanics.

The epigraph by Alexander Pope is an epitaph that he wrote for Isaac Newton, but which was not actually used on his grave.

CHAPTER 4

A textbook which also serves as an introduction to number theory is *Elementary Number Theory* by David M. Burton (Allyn and Bacon, London, 1976). An alternative introduction is provided by the intriguing mixture of information in *The Penguin Dictionary of Curious and Interesting Numbers* by David Wells (Penguin, Harmondsworth, 1986). A collection of problems in number theory which vary from serious research problems to problems suitable for amateurs is *Unsolved Problems in Number Theory* by Richard K. Guy (Springer-Verlag, New York, 1981). Some further superstitions about numbers can be gleaned from chapter 1 of *Number and Pattern in the Eighteenth-century Novel* by Douglas Brooks (Routledge and Kegan Paul, London, 1973). In pages 20–33 of *Invitation to Philosophy* by Martin Hollis (Basil Blackwell, Oxford, 1985), the difference between a proof of a proposition and evidence for the proposition is explained, and it is emphasized that proof is essential in mathematics. Pages 43–101 of E. R. Stabler's textbook *An Introduction to Mathematical Thought* (Addison-Wesley, Reading, Massachusetts, 1948) provide an introduction (not entirely elementary) to logic, and a more advanced introduction to logic is to be found in pages 109–71 of the textbook *The Foundations of Mathematics* by Ian Stewart and David Tall (Oxford University Press, Oxford, 1977).

The epigraph is poem number 3 in Diogenes O'Rell's *Poems for Clothilda* (Editions Clovis, Paris, 1966).

Stephen Leacock's short story 'A, B and C – The Human Element in Mathematics' is from his collection *Literary Lapses* (McClelland and Stewart, Toronto, 1980).

CHAPTER 5

Further information about congruences can be found in Ian Stewart's *Concepts of Modern Mathematics* (Pelican series paperback; Penguin, Harmondsworth, 1975) in pages 27–42 and 86–9.

CHAPTER 6

There is a good introduction to some of the ideas in this chapter in pages 65–88 of W. W. Sawyer's *Prelude to Mathematics* (Pelican series paperback; Penguin, Harmondsworth, 1955). Ian Stewart outlines the proofs of the impossibility of duplicating the cube and trisecting all angles by ruler and compass constructions in pages 81–6 of *Concepts of Modern Mathematics* (Pelican series paperback; Penguin, Harmondsworth, 1975). A more general introduction to geometry is H. S. M. Coxeter's fascinating textbook *Introduction to Geometry* (Wiley, New York, 1961).

There is a variety of problems like ours concerning bear hunting. For example, there is a similar bear problem in pages 15–16 of Eugene P. Northrop's amusing *Riddles in Mathematics* (Pelican series paperback; Penguin, Harmondsworth, 1960).

CHAPTER 7

Linear Equations by P. M. Cohn (Routledge and Kegan Paul, London, 1958) provides an excellent introduction to linear algebra, a subject for which there are many textbooks. *An Introduction to Matrices and Linear Transformations* by John H. Staib (Addison-Wesley, Reading, Massachusetts, 1969) is one textbook which covers the subject well but without unnecessary sophistication. Complex numbers are introduced in pages 216–33 of W. W. Sawyer's *Mathematician's Delight* (Pelican series paperback; Penguin, Harmondsworth, 1943), and a more formal account can be found in pages 208–15 of *The Foundations of Mathematics* by Ian Stewart and David Tall (Oxford University Press, Oxford, 1977). Abstract algebra is introduced in pages 76–112 of Ian Stewart's *Concepts of Modern Mathematics* (Pelican series paperback; Penguin, Harmondsworth, 1975). Unfortunately, I am precluded from recommending R. B. J. T. Allenby's textbook *Rings, Fields and Groups* (Edward Arnold, London, 1983) by the fact that the author cooperated in this book by criticizing its first draft, and this is a pity because he gives an interestingly gentle introduction to the subject starting from the material that is most familiar to the reader.

The poem from which the epigraph is taken appeared in *Linear Algebra and its Applications*, volume 41 (1981), pages 265–6.

Anyone who wishes to know more about Baron Ochs, Count Oktavian and the Field Marshall's wife (Princess Elizabeth von Werdenburg) should go to the nearest opera house which is presenting *Der Rosenkavalier* by Hugo von Hofmannsthal and Richard Strauss.

CHAPTER 8

Calculus is introduced in pages 105–76 of *Mathematician's Delight* by W. W. Sawyer (Pelican series paperback; Penguin, Harmondsworth, 1943). A textbook on calculus which is unusual for being designed for independent study is *Quick Calculus* by Daniel Kleppner and Norman Ramsey (Wiley, New York, 1965).

CHAPTER 9

A textbook on analysis which makes a special effort to alleviate the initial difficulties is *Mathematical Analysis* by R. Maude (Edward Arnold, London, 1986). An elementary textbook which aims to provide an easy introduction to the analysis of sequences and series is *A Text-book of Convergence* by W. L. Ferrar (Oxford University Press, Oxford, 1938).

The clerihew on John Napier is poem number 14 in Diogenes O'Rell's *Poems for Clothilda*.

CHAPTER 10

Men of Mathematics (two volumes in the Pelican series; Penguin, Harmondsworth, 1953) by E. T. Bell is a handy collection of lives of the great mathematicians, though it is something of a pantomime script with heroes to cheer and villains to hiss. The accounts of the views of Leibriz and Newton came from this book. W. W. Sawyer discusses the qualities needed by mathematicians and how these qualities relate to mathematics in pages 7–64 of *Prelude to Mathematics* (Pelican series paperback; Penguin, Harmondsworth, 1955). Edward de Bono launched his attack against excessive analysis in *The Use of Lateral Thinking* (Cape, London, 1969). W. W. Sawyer discusses methods of studying mathematics in pages 7–54 of *Mathematician's Delight* (Pelican series paperback; Penguin, Harmondsworth, 1943). That part of studying which involves the solving of problems is discussed in pages 285–98 of *The Mathematical Experience* by Philip J. Davis and Reuben Hersh (Pelican series paperback; Penguin, Harmondsworth, 1984). The best known textbook on problem solving is *How to Solve it* by G. Polya (Princeton University Press, princeton, 1945).

The epigraph was spoken by Laurence Sterne at a public dinner given in his honour.

The passage by William Hazlitt comes from his essay 'The Indian Jugglers' in *Table Talk*, from his complete works (Dent, London, 1931).

The origin of the Nobel Prizes is fully discussed in Lars Gårding and Lars Hörmander's paper 'Why is there no Nobel Prize in Mathematics?' in *The Mathematical Intelligencer*, volume 7 (1985), pages 73–4.

The poems by Diogenes O'Rell are numbered as follows in *Poems for Clothilda*: the haiku is 15, the limerick about competition is 92, 'The Leader Speaks' is 46 and Baron von Haerdupp's epitaph is 35.

The work of Issai Schur is outlined in the paper 'Issai Schur and his school in Berlin' by W. Ledermann in *The Bulletin of the London Mathematical Society*, volume 15 (1983), pages 97–106.

The paper 'The School of Hilbert and Emmy Noether' by B. L. van der Waerden, in *The Bulletin of the London Mathematical Society*, volume 15 (1983), pages 1–7, contains more details of the careers of David Hilbert and Emmy Noether.

The poem by Lewis Carroll concerning the disappointed papal candidate is from page 319 of *Sylvie and Bruno Concluded* (Macmillan, London, 1893), and it is the last stanza of 'The Mad Gardener's Song', which is scasttered throughout the two volumes of *Sylvie and Bruno*. Lewis Carroll never denied that the gardener's name was Martin.

Augustus De Morgan's comment on mathematicians in general is from page 151 of volume I of *A Budge of Paradoxes* (2nd edition, Open Court, Chicago, 1915).

The exposition of the Principle of Elimination occurs in 'The Blanched Soldier', which is in the collection *The Case-Book of Sherlock Holmes* by Sir Arthur Conan Doyle.

The view that music and mathematics produce a balance between meaning and emotion has also been advanced for music and science by Sir Brian Pippard in a music broadcast on BBC Radio 3 on 28 April 1979.

The quatrain from *The Rubaiyat of Omar Khayyam* that is quoted is number 69 in the first edition of Edward Fitzgerald's translation (republished by Gay and Hancock, London, 1917).

The comment by the Duchess in *Alice and Wonderland* by Lewis Carroll can almost be found in chapter 9.

The conversation between Sir Thomas More and Richard Rich is taken from Act 1 of Robert Bolt's *A Man for all Seasons*.

The problem about placing pigs into pens properly was inspired by a very similar problem in Knot VIII of Lewis Carroll's *A Tangled Tale* (republished by Dover, New York, 1958).

Index

Abel, N., 112, 179
acceleration, 39–41, 127
accuracy, 144
addition, 60, 116
aerodynamics, 36
a fortiori, 138
air resistance, 46–7
algebra, 3, 6, 32–3, 96–118, 139;
    abstract, 33, 107–17, 171;
    linear, 96–107, 117–18
algorist, 180
analysis, 3, 6, 139–59
Apollonius of Perga, 76–7, 80
approximation, 37, 87–8,
    124–5, 135, 143–4, 152, 155
Arabia, 3, 96,
arbitrary constant, 136
Archimedes, 3, 43, 131, 162,
    165
area, 131–4, 135
arithmetic, 2, 6, 15, 28–34,
    50–66, 139
Associative Law, 31, 115, 117
astronomy, 2, 35, 37, 38, 40–1,
    54–5, 171
autoreference, 196

axioms, 24, 25–49, 59, 81–2,
    84–6, 86, 117

Babbage, C., 28, 167
Babylon, 2, 3, 50, 51
Barrow, I., 165
Beaker Folk, 173–4
bear, 85, 95
Benz, F., 165
Bessel functions, 57
bison, 85
body, 37, 41
Boole, G., 4, 16, 28

Caesar, J., 10, 174
calculus, 4, 38–9, 119–38, 139,
    140–1, 186
Cambridge, 164
Cantor, G., 4, 16
Cardano, G., 3, 108, 109, 110
Cartesian coordinates, 80, 82–3,
    86
cat, 13–14
Cauchy, A., 4, 114
Cayley, A., 4, 86, 117, 118,
    166, 167

Index

Index